T0135625

Rüdiger Krause

Genetic Algorithms as Tool for Statistical Analysis of High-Dimensional Data Structures

Dissertation an der
Fakultät für Mathematik, Informatik und Statistik
der Ludwig-Maximilians-Universität München

Einreichung der Dissertation: 28.01.2004
Tag des Rigorosums: 25.05.2004

1. Gutachter: Prof. Dr. G. Tutz
2. Gutachter: Prof. Dr. L. Fahrmeir
3. Gutachter: Prof. Dr. I. Pigeot

Bibliografische Information Der Deutschen Bibliothek

Die Deutsche Bibliothek verzeichnet diese Publikation in der Deutschen
Nationalbibliografie; detaillierte bibliografische Daten sind im Internet über
http://dnb.ddb.de abrufbar.

ISBN 3-8325-0661-6

Logos Verlag Berlin
Comeniushof, Gubener Str. 47,
10243 Berlin
Tel.: +49 030 42 85 10 90
Fax: +49 030 42 85 10 92
INTERNET: http://www.logos-verlag.de

Acknowledgements

This thesis evolved from research on projects of the Department of Statistics and the Sonderforschungsbereich 386 "Statistical Analysis of Discrete Structures", Ludwig-Maximilians-University Munich.

The author thanks Gerhard Tutz, Department of Statistics, Ludwig-Maximilians-University Munich, for an fruitful cooperation with extensive consultation and support at all stages of the scientific projects.

I would also like to thank Ludwig Fahrmeir, Department of Statistics, Ludwig-Maximilians-University Munich, and Iris Pigeot, Department of Statistics, University of Bremen, for providing written expertises and detailed reviews.

I am particularly grateful to the following people for careful review and valuable discussions: Christiane Belitz, Angelika Blauth, Anne-Laure Boulesteix, Jochen Einbeck, Eva-Maria Fronk, Klaus Hechenbichler, Alexander Jerak, Stefan Lang, Florian Leitenstorfer and Torsten Scholz.

Furthermore I wish to thank colleagues, students and staff at the Department of Statistics and the Sonderforschungsbereich 386 for the familiar working atmosphere.

Munich,
May 2004

Rüdiger Krause

Contents

Part II Genetic Algorithms for Smoothing Parameter Choice

1

Introduction

The basic interest in statistical modelling is the analysis and interpretation of a given data set by application of statistical tools or models. Ideally the statistician would like to determine a function which reflects reality as accurate as possible but also eliminates irregularities from data noise and is therefore easy to interpret. For example in regression the objective is to determine an appropriate function which adequately describes the available data set but also contains the relevant information of the data.

All the observed data include noise resulting from e.g. measurement errors or disturbing influences. Therefore the user has always the problem to decide which information can be explained by the statistical model and which information is irrelevant and can be ignored by the model.

An approach which flexibly approximates the unknown function is the expansion in basis functions (see e.g. Fahrmeir & Tutz (2001)). Here the function is represented by a linear combination of finite basis functions.

A problem of this approach is to determine the appropriate number of basis functions. A large number of basis functions yields function estimators which are close to the observed data. Hence the run of the curves is wiggly and difficult to interprete. In contrast small numbers of chosen basis functions yield an opposite result: very smooth function estimators, which badly reproduce the observed data and thus the hidden data information is almost completely lost.

Two different approaches are generally used to solve this dilemma: *smoothing methods* and *adaptive methods*. Smoothing methods (Ruppert, Wand & Carroll (2003), Hastie, Tibshirani & Friedman (2001)) expand the function to be estimated as linear combination of many basis functions. Addition of a penalization term causes a tradeoff between data adaption (because of the large number of basis functions) and smoothness of the estimator function. In contrast adaptive methods avoid the penalization term by expanding the function to be estimated in a linear combination containing less basis functions. The number of basis functions used is decided by adaptive selection procedures (Miller (2002)).

Beside function estimation, statisticans often have to solve another important problem: in many statistical applications (e.g. analysis of gene expression data) there are large sets of explanatory variables containing many redundant or irrelevant variables. In recent years computers with better computing power have been developed and hence data sets with an increasing number of variables can be analysed. However the user is more frequently confronted with numbers of explanatory variables so large that present computation techniques are still difficult to apply. Thus adequate *variable selection* procedures (Miller (2002)) are necessary to shrink the number of relevant variables.

In many statistical applications the user has the problem of choosing an adequate subset from a large set of variables and simultaneously get appropriate estimators. Some examples are: variable selection with smoothing parameter choice or variable selection connected with simultaneous choice of knot locations. Generally software packages offer separate tools which are applied successively. The disadvantage of separate tools can be described in the following way: if the user is interested e.g. in a variable selection some parameters (e.g. smoothing parameters) have to be chosen previously. Usually these "roughly" chosen parameters are unchanged during variable selection. A fine tuning of the parameters is subsequently done by another tool. The problem is that a chosen variable subset is generally optimal in respect to the default parameters; however other parameters can cause different variable subsets. Thus a subsequent choice of parameters by another software tool can often lead to limited improvements, only. An approach which selects variables and parameters simultaneously should yield significantly improved results.

The literature provides a large number of approaches respectively algorithms which deal with detection of appropriate sets of redundant variables respectively adequate function estimation. A detailed presentation is given e.g. in Ruppert, Wand & Carroll (2003), Hastie, Tibshirani & Friedman (2001) and Miller (2002) (in this thesis we will present some approaches in context with the simulation studies). The performance of the diverse approaches depends on the given problem and is particularly influenced by prediction accuracy of the true underlying function as well as computational cost.

In contrast to the methods mentioned above *genetic algorithms* are rarely applied in the field of statistics. Genetic algorithms are more popular in computer science or optimization. They refer to the principle that better adapted (more fit) individuals win against their competitors under equal conditions. Like their biological standard these algorithms use biological components like selection, crossover, or mutation to model the natural phenomenon of genetic inheritance and Darwinion strife of survival.

Genetic algorithms have the disadvantage to be a computationally expensive method compared with many common approaches. But these approaches often yield inappropriate solutions for many existing complex problems (e.g. optimization problems which have several local optima apart from the global optimum). Genetic algorithms often yield better results for complex problems because they do not require knowledge or gradient information about the response surface. Other advantages of genetic algorithms is the possibility to independently jump out of a given local optimum and the ability to search the search space at several locations simultaneously. Genetic algorithms always yield a list of several solutions. Each other approach leads to one solution (i.e. only one point in the search space).

In this thesis genetic algorithms are applied to several interesting statistical problems, e.g.

- appropriate choice of smoothing parameters in smoothing methods;

- adequate subset of variables;

- simultaneous selection of variables and parameters (e.g. smoothing parameters).

The new presented genetic algorithms in this thesis are universelly applicable i.e. they are applicable to all problems related with selection or optimization of variables and parameters. Some examples are knot selection in adaptive approaches, optimization of characteristical parameters in basis functions (e.g. choice of the spread in Gaussian kernels) or the choice of adequate parameters in algorithms. In this thesis however genetic algorithms are essentially restricted to examples of smoothing parameters and variable selection.

This thesis is divided into four main parts, each one containing several chapters.

The *first part* introduces diverse basics of statistical modelling and multidimensional regression. In detail this part starts with a presentation of additive models which are a flexible approach for estimating the true underlying regression function (section 2.1). Section 2.2 describes the additive logistic model which is popular in biometrics. The following section deals with the popular approach of expanding the functions (e.g of an additive model) in basis functions. Furthermore we describe some types of basis functions commonly used, e.g. radial basis functions or splines. Chapter 3 presents the concept of estimation and penalization as applied to smoothing methods. Here the penalization term is adapted to the additive model used in the thesis. The last chapter of the first part provides a short survey of diverse information criteria (e.g. AIC, GCV) which are interesting tools for adequate model selection.

The *second part* introduces genetic algorithms. Beside "classical" binary coded genetic algorithms also real-coded ones are described. In chapter 5 we present a new modified genetic algorithm for optimization of continuous parameters (GENcon). For a better performance diverse new operators (e.g. improved arithmetical crossover operator, section 5.3.2) are developed. As a special case we apply the algorithm to the problem of smoothing parameter choice. A simulation study (chapter 6) compares our approach with other methods presented in literature. Finally in chapter 7 we discuss the question which default parameter combination of the genetic algorithm may be an optimal choice. To answer this question diverse simulations are considered.

The *third part* deals with the problem of variable selection. As solution method we suggest a genetic algorithm for binary values (GENbin) in chapter 8. As the algorithm differs from GENcon in the coding we also have to adapt the biological components (chapter 8). Chapter 9 uses simulations to compare our approach with other common methods in literature. In chapter 10 the genetic algorithm for variable selection is applied to a real dataset based on gene expression data and hence to a problem of classification.

Finally in *part four* we construct a genetic algorithm for simultaneous selection of continuous parameters (e.g. smoothing parameters) and variables. We formulate the algorithm in such a way that each selection problem (i.e. not only variable selection) can be considered. Therefore the components of the genetic algorithms GENcon and GENbin are combined and adapted for the case of simultaneous selection (chapter 11). In chapter 12 an extended simulation study assists to compare our approach with other methods for the special case of simultaneous selection of variables and smoothing parameters. In chapter 13 we apply the genetic algorithm for simultaneous selection of variables and parameters to a real dataset, the "rental guide" of Munich.

The *appendix* of this thesis contains the manuals of the software packages which are applied in context with the diverse simulation studies. The software can be downloaded from internet. The address is *http://www.stat.uni-muenchen.de/sfb386*.

Part I

Additive Modelling with Penalized
Regression Splines

Expansion of Additive Models in Basis Functions

2.1 Additive Models

Suppose we have a dataset consisting of observations $(y_i, \mathbf{x}_i), i = 1, \ldots, n,$ and each $\mathbf{x}_i$ being a vector of p components $\mathbf{x}_i = (x_{i1}, \ldots, x_{ip})$. One of the most popular statistical tools is the regression model. Here the response variable y_i depends on the covariate $\mathbf{x}_i$ by a relation

$$\begin{aligned} y_i &= f(\mathbf{x}_i) + \epsilon_i \\ &= f(x_{i1}, \ldots, x_{ip}) + \epsilon_i \quad , \end{aligned} \tag{2.1}$$

where $f(.)$ is an unspecified smooth function. The error ϵ_i is assumed to be independent of $\mathbf{x}_i$ and to have normal distribution with zero mean, i.e. $\epsilon_i \sim \mathcal{N}(0, \sigma^2)$.

In statistical modelling an essential objective is the search for a suitable estimator $\hat{f}$ which appropriately describes the usual noisy observations. In this context a fundamental problem is to define what an appropriate approximation of a function means. If we choose f in (2.1) in a way that the data are described as precisely as possible, we have the extreme case of interpolation. Here each covariate $\mathbf{x}_i$ is mapped exactly to the corresponding response variable

$$f(\mathbf{x}_i) = y_i \quad , \qquad i = 1, \ldots, n, \tag{2.2}$$

(Figure 2.1(a)). However most problems (e.g. problems of diagnostics and forecast) require detection of the structure which is hidden in the data. Smoothing of the approximate function $\hat{f}$ by diverse smoothing methods (Hastie, Tibshirani & Friedman (2001)) reduces the number of wiggles

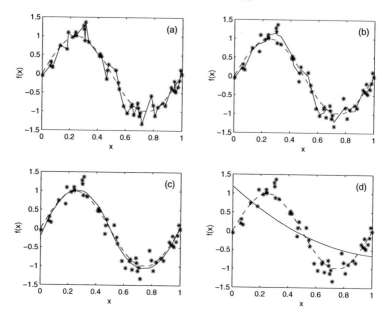

Figure 2.1. *A set of 50 data points is generated by sampling a function $f(x) = sin(6.3x)$, shown by the dashed curve. Gaussian noise with a standard deviation 0.2 is added. Based on the 50 data points an estimatior $\hat{f}$ is calculated (solid curve). Subplot (a) illustrates the case of interpolation while subplots (b)-(d) are examples of increasing smoothing, with (d) as an extreme case of an underfitted function. An appropriate solution for this example is given in subplot (c).*

and detects hidden structure (Figure 2.1(b) and (c)). One has to pay attention, however, that too much smoothing can blur the true character and the curvature of the function. In this case we have an underfitted function (Figure 2.1(d)).

A popular and flexible approach for estimating the true underlying regression function, which assumes some structure in the predictor space, is the *additive model* discussed in detail by Hastie & Tibshirani (1990). In the additive model it is assumed that the response variable y_i depends on $\mathbf{x}_i$ by

$$y_i = \beta_0 + f_1(x_{i1}) + \ldots + f_p(x_{ip}) + \epsilon_i$$
$$= \beta_0 + \sum_{j=1}^{p} f_j(x_{ij}) + \epsilon_i \quad , \tag{2.3}$$

where $\epsilon_i \sim \mathcal{N}(0, \sigma^2)$ and $f_1, \ldots, f_p$ are unknown smooth functions which have to be estimated. It is obvious that the additive model replaces the problem of estimating a function f of a p-dimensional variable $\mathbf{x}_i$ by one of estimating p separate one-dimensional functions $f_j(x_{ij})$. The advantage of (2.3) is its potential as a data analytic tool: since each variable is represented separately one can plot the p co-ordinate functions separately and thus evaluate the roles of the single predictors. With the assumption that the response variable y_i is not normally distributed, (2.3) can be extended to the *generalized additive model* (Hastie & Tibshirani (1990)) which has the form

$$y_i = g(\eta_i) + \epsilon_i$$
$$= g\{\beta_0 + \sum_{j=1}^{p} f_j(x_{ij})\} + \epsilon_i \quad , \tag{2.4}$$

where $\eta_i = \beta_0 + f_1(x_{i1}) + \ldots + f_p(x_{ip})$ is an additive predictor and $g(.)$ is a known link or response function. By choosing $g(.)$ as identity function the generalized additive model again reduces to an additive model.

The additive model in (2.3) is not necessarily restricted to component functions with one-dimensional metrical variables. Usually additive models have interactions between two or more metrical variables and then the additive model in (2.3) can be expanded in

$$y_i = \beta_0 + \sum_{j=1}^{p} f_j(x_{ij}) + \sum_{r=1}^{p-1} \sum_{s=r+1}^{p} f_{rs}(x_{ir}, x_{is}) + \epsilon_i \tag{2.5}$$

where $\epsilon_i \sim \mathcal{N}(0, \sigma^2)$ and the term with the double sum describes the interaction between different metrical variables. It should be noticed that not all one-dimensional metrical variables must necessarily occur in the interaction terms. Furthermore we restrict ourselves to interaction terms between two variables.

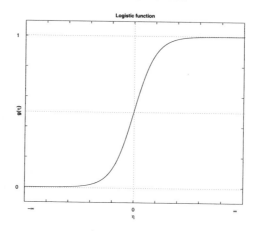

Figure 2.2. *Here is shown the logistic function $g(\eta) = 1/(1 + exp(-\eta))$. The range of $g(\eta)$ is between 0 and 1. The logistic function $g(\eta)$ equals 0 if $\eta = -\infty$ and $g(\eta)$ equals 1 if $\eta = \infty$.*

Remark 1. Additive models can be extended to models with categorical variables and their interactions with metrical variables. We will deal with these extended additive models in section 3.3 in detail.

2.2 Additive Logistic Model

A special case of the generalized additive model in (2.4) is the *additive logistic model* (Hastie, Tibshirani & Friedman (2001)). Supposed we have observed independent variables $\mathbf{x}_i = (x_{i1}, \ldots, x_{ip}), i = 1, \ldots, n$, which are connected with dichotomous dependent variables $y_i, i = 1, \ldots, n$ (e.g. $y_i \in \{0, 1\}$ can characterize existence of tumor with $y_i = 0$ (no tumor) and $y_i = 1$ (tumor)). Now we are interested in analysing the probability for given $\mathbf{x}_i$ to get the event $A(y_i = 1)$ (e.g. patient has tumor), i.e.

$$E(y_i|\mathbf{x}_i) = P(y_i = 1|\mathbf{x}_i) = \pi(\mathbf{x}_i).$$

If we insert in (2.4) the logistic function $g(\eta) = 1/(1 + exp(-\eta))$ as link function g (compare Figure 2.2) we receive the additive logistic model

$$\pi(\mathbf{x}_i) = \frac{1}{1 + e^{-(\beta_0 + \sum_{j=1}^{p} f_j(x_{ij}))}} \quad . \tag{2.6}$$

In case we are restricted to a linear model $y_i = \beta_0 + \beta_1 x_{i1} + \ldots + \beta_p x_{ip} + \epsilon_i$ the formula (2.6) can be written as the *logistic model* (or *logit-model*)

$$\pi(\mathbf{x}_i) = \frac{1}{1 + e^{-(\beta_0 + \sum_{j=1}^{p} \beta_j x_{ij})}} \quad , \tag{2.7}$$

where $\beta_0, \beta_1, \ldots, \beta_p$ are the parameters which can be estimated with the maximum-likelihood method (chapter 10). If $\mathbf{x}_i$ is a metrical value (as used e.g. in (2.7)) we have *logistic regression.*

2.3 Expansion in Basis Functions

An approach which allows flexible representations of the functions $f_j(x_{ij})$, $j = 1, \ldots, p$, in (2.5) is the *expansion in basis functions*, i.e. $f_j(x_{ij})$ is represented as

$$f_j(x_{ij}) = \sum_{\nu=1}^{K_j} \beta_{j\nu} \, \phi_{j\nu}(x_{ij}) \tag{2.8}$$

where $\beta_{j\nu}$ are unknown coefficients and $\{\phi_{j\nu}(x_{ij}), \nu = 1, \ldots, K_j\}$ is a set of basis functions. Thus the function $f_j(x_{ij})$ is represented by a linear combination of finite basis functions. Each basis function $\phi_{j\nu}(x_{ij})$ is characterized by a knot $\xi_{j\nu}$ from the range of the jth covariate. In context with additive models the typically non-linear basis functions $\phi_{j\nu}$ can be considered as a transformation

$$\phi_{j\nu} : \mathbb{R} \longrightarrow \mathbb{R}$$
$$x_{ij} \mapsto \phi_{j\nu}(x_{ij}).$$

The advantage of this approach is its simple estimation: if we substitute the diverse basis functions by new variables $(z_{ij\nu} \equiv \phi_{j\nu}(x_{ij}), \nu = 1, \ldots, K_j)$ we have a linear model, i.e.

$$f_j(x_{ij}) = \beta_{j1} z_{ij1} + \ldots + \beta_{jK_j} z_{ijK_j}$$

and then we can apply the common least squares method (in case of normal distribution) for estimating.

The approach obtains its flexibility from the diverse possible choices of numbers and types of basis function. In the literature there are several popular types of basis functions, e.g. radial basis functions and splines. Both groups of basis functions are described in the following subsections.

The interaction term $f_{rs}(x_{ir}, x_{is})$ in (2.5) can also be expanded in basis functions. In this case the two-dimensional function $f_{rs}(x_{ir}, x_{is})$ is represented as a tensor product of two one-dimensional basis functions, i.e.

$$f_{rs}(x_{ir}, x_{is}) = \sum_{\kappa=1}^{K_r} \sum_{\rho=1}^{K_s} \gamma_{rs,\kappa\rho}\, \phi_{r\kappa}(x_{ir})\phi_{s\rho}(x_{is}) \quad , \tag{2.9}$$

where the numbers of basis functions K_r, K_s for the two metrical variables can be unequal.

2.3.1 Radial Basis Functions

Radial basis functions are characterized by knots $\boldsymbol{\xi}_j = (\xi_{j1}, \ldots, \xi_{jK_j}), j = 1, \ldots, p$, determining the centres of the basis functions $\phi_{j\nu}(x_{ij}), \nu = 1, \ldots, K_j$. Here radial functions are used as basis functions. They have the property that the response variable y_i decreases or increases monotonically with the distance from the central points $\boldsymbol{\xi}_j, j = 1, \ldots, p$. In this context $\phi_{j\nu}(.)$ is some non-linear function depending on the (usually) Euclidean distance between x_{ij} and $\xi_{j\nu}$, i.e. we have terms of the form $\phi_{j\nu}(|x_{ij} - \xi_{j\nu}|)$. Some examples of radial basis functions are (compare Figure 2.3)

- $\phi_{j\nu}(x_{ij}) = exp\left(-\frac{|x_{ij}-\xi_{j\nu}|^2}{2\sigma_{j\nu}{}^2}\right)$ (Gaussian kernel),

- $\phi_{j\nu}(x_{ij}) = (|x_{ij} - \xi_{j\nu}|^2 + \sigma_{j\nu}{}^2)^{-1}$ (inverse multi-quadratic function),

- $\phi_{j\nu}(x_{ij}) = |x_{ij} - \xi_{j\nu}|^2 \, ln\,(|x_{ij} - \xi_{j\nu}|)$ (thin-plate spline function),

- $\phi_{j\nu}(x_{ij}) = (|x_{ij} - \xi_{j\nu}|^2 + \sigma_{j\nu}{}^2)^{-\frac{1}{2}}$ (multi-quadratic function),

- $\phi_{j\nu}(x_{ij}) = (|x_{ij} - \xi_{j\nu}|)^3$ (cubic function),

- $\phi_{j\nu}(x_{ij}) = |x_{ij} - \xi_{j\nu}|$ (linear function).

Here $\sigma_{j\nu}, j = 1, \ldots, p, \nu = 1, \ldots, K_j$ is a parameter which controls the spread of the basis functions. Gaussian kernel- and inverse multi-quadric basis functions have the property $\phi_{j\nu}(x_{ij}) \longrightarrow 0$ as $|x_{ij}| \longrightarrow \infty$ and are denoted as *localized* basis functions. The other examples, called *non-localized* basis functions, are characterized by $\phi_{j\nu}(x_{ij}) \longrightarrow \infty$ as $x_{ij} \longrightarrow \infty$.

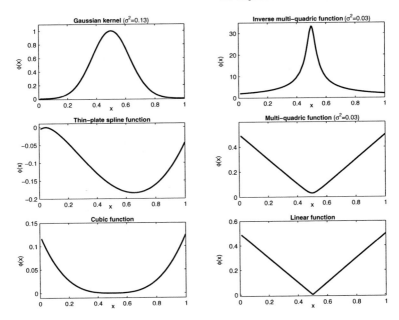

Figure 2.3. *Some examples of radial basis functions for one covariate x with a knot $\xi = 0.5$. The subplots in the first row are localized functions, the other ones are non-localized functions.*

2.3.2 Splines

A spline can be described in a simplified way as defining a function piecewise. Then, these single pieces are connected at the separate points with certain smoothing properties. An exact definition of splines is given below. Because of simplicity in the next three sections we only consider (without any loss of generality) additive models consisting of one component, i.e. $f(x) \equiv f_1(x_{i1})$ and thus the indices i and j can be omitted.

Definition 1. *Let $S_\nu = \{\xi_0, \ldots, \xi_K\}$ be a system of knots strictly increasing on a finite interval $[a, b] \subset \mathbb{R}$, i.e.*

$$a = \xi_0 < \xi_1 < \ldots < \xi_K = b \ .$$

We call a function $s_d : [a, b] \to \mathbb{R}$ spline function of degree $d = 0, 1, \ldots$ (and order $d + 1$), if the following conditions are satisfied:

(i) The function $s_d(x), x \in \mathbb{R}$, and its derivatives up to order $d - 1$ are all continuous on $[a, b]$, i.e.

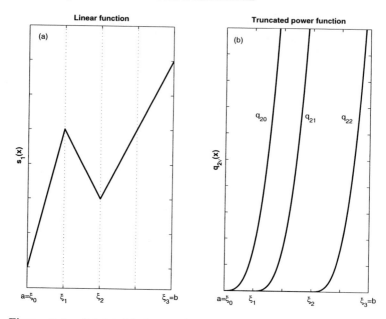

Figure 2.4. *Subplot (a) shows a linear spline $s_1(x) \in s_1(S_4)$ of degree=1 and order=2. In (b) there is plotted a truncated power function of degree=2, i.e. $q_{2\nu}(x) \in s_2(S_4), \nu = 0, 1, 2$. Here $s_1(S_4)$ is a system of four knots strictly increasing on an interval $[a, b]$.*

$$s_d(x) \in C^{d-1}([a, b]). \tag{2.10}$$

(ii) $s_d(x)$ is for $x \in [\xi_\nu, \xi_{\nu+1}], \nu = 1, 2, \ldots, K - 1$ a polynomial of degree d at most

$$s_d(x) \mid_{[\xi_\nu, \xi_{\nu+1}]} \in P_d \quad , \nu = 0, 1, \ldots, K - 1. \tag{2.11}$$

For the partition S_ν we call the set of all splines of degree d $s_d(S_\nu)$.

Remark 2. In the following we consider only those splines, which are put together piecewise by polynomials. Hence, if we talk about splines (as we did in Definition 1), we always think of polynomial-splines.

Each polynomial of degree d is a spline of $s_d(S_\nu)$ with some partition S_ν, while not every spline is a polynomial. Simple examples of spline functions are the linear spline $s_1(S_\nu)$ or the *truncated power function* (compare Figure 2.4). Given a partition S_ν, a linear spline consists of a frequency polygon, which connects the single knots linearly. On the other side the truncated power function

$$q_{d\nu} : [a, b] \longrightarrow \mathbb{R}, \ 0 \leq \nu \leq K - 1$$

$$q_{d\nu}(x) = (x - \xi_\nu)_+^d = \begin{cases} (x - \xi_\nu)^d & for \quad x \geq \xi_\nu \\ 0 & for \quad x < \xi_\nu \end{cases} \qquad (2.12)$$

is a spline of degree d for partition S_ν. However, $q_{\nu1}, \ldots, q_{d,K-1}$ are not polynomials of $[a, b]$. From Definition 1 we realize that the sets $s_d(S_\nu)$ are linear subspaces of $C^{d-1}[a, b]$ and therefore $s_d(S_\nu)$ are vector spaces. The following Theorem 1 presents one possible basis of $s_d(S_\nu)$:

Theorem 1. *Each spline $s_d(x) \in s_d(S_\nu)$ has a unique representation*

$$s_d(x) = \sum_{i=0}^{d} a_i x^i + \sum_{\nu=1}^{K-1} c_\nu (x - \xi_\nu)_+^d. \qquad (2.13)$$

The monoms x^i and truncated power functions form a basis

$$\mathcal{B} := \{1, x, \ldots, x^d, (x - \xi_1)_+^d, \ldots, (x - \xi_{K-1})_+^d\}$$

of the vector space $s_d(S_\nu)$. Especially the vector space is of dimension $(d + K)$.

Proof. We prove this theorem in two steps: in (i) we show, that it is allowed to use $(d + K)$ free parameters at maximum for the construction of a spline $s_d(S_\nu)$. In step (ii) we show that $(d + K)$ functions in $\mathcal{B}$ are linearly independent and hence the vector space $s_d(S_\nu)$ is of dimension $(d + K)$.

(i) For an interval $[\xi_0, \xi_1]$ we can select every spline of degree $\leq d$, which implies $(d + 1)$ free parameters. The smoothing property (2.10) leads to the condition

$$c_{\nu-1}(x - \xi_\nu)^d = c_\nu(x - \xi_\nu)^d, \quad \xi_\nu \leq x \leq \xi_{\nu+1}, \ \nu = 1, \ldots, K - 1.$$

Based on this condition the polynomials on the subsequent intervals $[\xi_1, \xi_2], \ldots, [\xi_{K-1}, \xi_K]$ are all determined -except of one parameter- by their predecessor.

(ii) Let

$$s_d(x) = \sum_{i=0}^{d} a_i x^i + \sum_{\nu=1}^{K-1} c_\nu (x - \xi_\nu)_+^d = 0 \quad \text{for all } x \in [a, b].$$

Consider the linear functionals

$$G_\nu(f) := \frac{1}{d!}(f^{(d)}(\xi_\nu^+) - f^{(d)}(\xi_\nu^-))$$

where $f(\xi_\nu^+)$ and $f(\xi_\nu^-)$ are the right- and left-sided limiting values respectively. If we apply the functionals to s_d, we get for all $\nu = 1, \ldots, K-1$

$$
\begin{aligned}
0 &= G_\nu(s_d) \\
&= \underbrace{G_\nu(\sum_{j=0}^{d} a_j x^j)}_{=0} + \sum_{j=1}^{K-1} c_j \underbrace{G_\nu(x - \xi_j)_+^d}_{=\delta_{\nu j}} \\
&\stackrel{(2.12)}{=} \frac{1}{d!} c_\nu \left[\underbrace{\left((x - \xi_\nu^+)_+^d\right)^{(d)}}_{=d!} - \underbrace{\left((x - \xi_\nu^-)_+^d\right)^{(d)}}_{=0} \right] \\
&= c_\nu
\end{aligned}
$$

where $\delta_{\nu j} = \begin{cases} 1 & \text{if } j = \nu \\ 0 & else \end{cases}$ is the Kronecker symbol. That yields

$$s_d(x) = \sum_{i=0}^{d} a_i x^i = 0 \quad \text{for all } x \in [a, b],$$

and this implies that $a_0 = a_1 = \ldots = a_d = 0$. Hence the $(d + K)$ functions in $\mathcal{B}$ are linearly independent and the vector space $s_d(S_\nu)$ is of dimension $(d + K)$. $\qquad\square$

However the basis $\mathcal{B}$ of $s_d(S_\nu)$ has some disadvantages:

- The elements of $\mathcal{B}$ are not local because the support of the monomes x^i is for example the whole space $\mathbb{R}$.

- If two knots $\xi_\nu, \xi_{\nu+1}$ are close together, the truncated power series basis becomes "almost" linearly dependent. Therefore the spline in representation (2.13) is ill-conditioned with reference to disturbances in the coefficients c_ν. Hence the basis $\mathcal{B}$ is unsuitable for numerical calculations.

- The coefficients a_i and c_ν in (2.13) cannot be interpreted geometrically.

2.3.3 B(asic-) splines

In this section we describe the B(asic-) spline basis, which has better numerical properties compared with the truncated power series basis. A detailed presentation about B-splines is given in de Boor (1978) or de Boor (1993) and Dierckx (1995).

Definition 2. *Let* $\xi_1 \leq \ldots \leq \xi_K$ *be any sequence of knots. A B-spline* $B_{d,\nu}(x)$ *of degree* $d \geq 0$ *which starts at knot* ξ_ν *is defined by*

$$B_{0,\nu}(x) = \begin{cases} 1 & if \ \ \xi_\nu \leq x \leq \xi_{\nu+1} \\ 0 & else \end{cases} \tag{2.14}$$

$$B_{d,\nu}(x) = \frac{x - \xi_\nu}{\xi_{\nu+d} - \xi_\nu} B_{d-1,\nu}(x) + \frac{\xi_{\nu+d+1} - x}{\xi_{\nu+d+1} - \xi_{\nu+1}} B_{d-1,\nu+1}(x) \ . \tag{2.15}$$

Remark 3. For equidistant knots which are used here (2.15) simplifies to

$$B_{d,\nu}(x) = \frac{1}{d \cdot d\xi} \left[(x - \xi_\nu) B_{d-1,\nu}(x) + (\xi_{\nu+d+1} - x) B_{d-1,\nu+1}(x) \right]$$

because $\xi_{\nu+d} - \xi_\nu = \xi_{\nu+d+1} - \xi_{\nu+1} = d \cdot d\xi$ where $d\xi$ is the distance between two adjacent knots.

Figure 2.5(a) shows B-splines of degree 1, respectively order 2. Here at each knot $\xi_\nu, \nu = 1, \ldots, K$, a B-spline is generated by joining of two polynomials of degree 1 (piecewise linear functions). Figure 2.5(b) shows B-splines of degree 2 and order 3 (quadratic B-splines). At the inner knots (for example $\xi_{\nu+1}$ and $\xi_{\nu+2}$) three polynomials of degree 2 are connected. For this kind of B-splines the first derivatives are equal at the joining points. This does not hold for the second derivatives. Most commonly B-splines of degree 3 respectively order 4 (cubic B-splines) are used. This type of B-splines are generated by four polynomials of degree 3 (Figure 2.5(c)) where the polynomials are connected at the inner knots. In this case the first and the second derivatives are equal at the junctures.

B-splines of degree d have the following general properties:

- B-splines consist of $d + 1$ polynomial pieces, each of degree d;

- they have d inner knots where the polynomial pieces are connected;

- B-splines have an overlap of $2d$ neighbouring B-splines. Of course the leftmost and the rightmost B-splines have less overlap;

- at the junctures, derivatives up to order $d - 1$ are continuous;

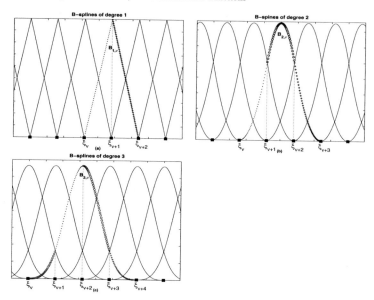

Figure 2.5. *Here B-splines of degree 1, 2, and 3 are shown. In each panel the different polynomials of one B-spline are exemplarily plotted.*

- B-splines are positive on a domain spanned by $d + 2$ knots; outside of this domain the B-spline is zero.

Note that each interval between two adjacent knots is covered by $d + 1$ B-splines of degree d. The basis functions $\phi_{j\nu}$ depend on one knot only. When using one knot to identify a specific B-spline we take the leftmost knot at which the spline becomes non-zero.

Beside numerical stability some other B-spline properties are interesting for users:

- $B_{d,\nu}(x) \geq 0$ for all $x \in \mathbb{R}$.

- $B_{d,\nu}(x) > 0$ for all $x \in (\xi_\nu, \xi_{\nu+d+1})$, i.e. $B_{d,\nu}$ has a local support.

- For all $x \in [a, b]$ we have

$$1 = \sum_{\nu=1}^{K} B_{d,\nu}(x),$$

i.e. B-splines have a decomposition of one on an interval $[a, b]$.

Remark 4. B-splines $B_{d,\nu}$ are splines of degree d and form a basis in the spline space (see e.g. de Boor (1978)).

2.3.4 Multidimensional B-splines

In this section we generalize the concept of one-dimensional B-splines to higher dimensions. For calculation of two-dimensional B-splines we can use the following Definition 3:

Definition 3. *Let $\xi_1 \leq \ldots \leq \xi_{K_x}$ and $\zeta_1 \leq \ldots \leq \zeta_{K_z}$ be any sequences of knots for two metrical variables x and z. A two-dimensional B-spline $B_{d,\nu\kappa}(x,z)$ of degree $d \geq 0$ which starts at the pair of knots (ξ_ν, ζ_κ) is defined by*

$$B_{0,\nu\kappa}(x,z) = \begin{cases} 1 & if \quad \xi_\nu \leq x \leq \xi_{\nu+1} \wedge \zeta_\kappa \leq z \leq \zeta_{\kappa+1} \\ 0 & else \end{cases} \qquad (2.16)$$

$$
\begin{aligned}
B_{d,\nu\kappa}(x,z) = \quad & \frac{(\xi_{\nu+d+1} - x)(\zeta_{\kappa+d+1} - z)}{(\xi_{\nu+d+1} - \xi_{\nu+1})(\zeta_{\kappa+d+1} - \zeta_{\kappa+1})} B^x_{d-1,\nu+1}(x) B^z_{d-1,\kappa+1}(z) + \\
& + \frac{(\xi_{\nu+d+1} - x)(z - \zeta_\kappa)}{(\xi_{\nu+d+1} - \xi_{\nu+1})(\zeta_{\kappa+d} - \zeta_\kappa)} B^x_{d-1,\nu+1}(x) B^z_{d-1,\kappa}(z) + \\
& + \frac{(x - \xi_\nu)(\zeta_{\kappa+d+1} - z)}{(\xi_{\nu+d} - \xi_\nu)(\zeta_{\kappa+d+1} - \zeta_{\kappa+1})} B^x_{d-1,\nu}(x) B^z_{d-1,\kappa+1}(z) + \\
& + \frac{(x - \xi_\nu)(z - \zeta_\kappa)}{(\xi_{\nu+d} - \xi_\nu)(\zeta_{\kappa+d} - \zeta_\kappa)} B^x_{d-1,\nu}(x) B^z_{d-1,\kappa}(z)
\end{aligned}
\qquad (2.17)
$$

Remark 5. Similar to (2.9) for basis functions, also two-dimensional B-splines $B_{d,\nu\kappa}(x,z)$ are calculated by a tensor product of two one-dimensional B-splines $B^x_{d,\nu}(x)$ in x-direction and $B^z_{d,\kappa}(z)$ in z-direction.

Figure 2.6 shows two-dimensional B-splines $B_{d,\nu\kappa}(x,z)$ of degree $d = 1, 2, 3$. For illustration we have also plotted the respective one-dimensional B-splines $B^x_{d,\nu}(x)$ and $B^z_{d,\kappa}(z)$, which generate $B_{d,\nu\kappa}(x,z)$ by calculating formula (2.17). From Figure 2.6 we realize that two-dimensional B-splines of e.g. degree $d = 3$ necessarily depend on five knots in each variable direction. The projected one-dimensional cubic B-splines are comparable with the corresponding ones in Figure 2.5.

A generalization of the B-spline concept to more than two dimensions is possible. But for increasing dimensions the B-spline basis grows exponentially fast. That means if we assume that K B-splines are given in each direction, the number of B-splines in a d-dimensional space is K^d (e.g. for $K = 3$ we get 9 B-splines in a two-dimensional space respectively 27 B-splines in a cubic space).

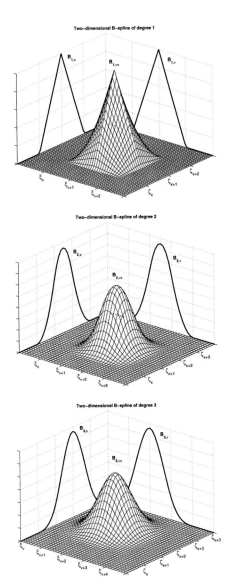

Figure 2.6. *Here two-dimensional B-splines $B_{d,\nu\kappa}(x,z)$ of degree $d = 1, 2, 3$ are shown. Furthermore we have plotted the two one-dimensional B-splines $B_{d,\nu}^x(x)$ and $B_{d,\nu}^z(z)$ which generate $B_{d,\nu\kappa}(x,z)$ by a tensor product.*

3

Estimation with Penalized Shrinkage

In chapter 2.3 we described the approach of "expansion in basis functions", which allows flexible approximations of a true unknown function. Here the estimator function is represented as a linear combination of basis functions. A substantial problem is the adequate number of basis functions. A number of basis functions being too large yields function estimators which are close to the observed data. Hence the run of the curves are wiggly and badly interpretable. On the other hand small numbers of chosen basis functions yield an oppositional result: very smooth function estimators, which reproduce the observed data badly and thus the hidden data information is almost completely lost. For solving this problem smoothing methods are commonly used (Hastie, Tibshirani & Friedman (2001)). These approaches expand the estimator function as linear combination of many basis functions. By addition of a penalization term there is a tradeoff between data adaption and smoothness of the estimator function.

This chapter is organized as follows: the next section describes the concept of estimation and penalization for additive models with metrical variables but without interactions. Then section 3.2 treats of the more general case of additive models with metrical variables and their interactions. Finally section 3.3 extends the additive model to categorical variables and their interactions with metrical variables. Here we discuss how estimation of terms depending on categorical variables can be integrated in the estimation concept for additive models.

3.1 Penalized Regression Splines without Interactions

In case of an additive model without interactions which has response variable y_i and metrical variables $\mathbf{x}_i = (x_{i1}, \ldots, x_{ip})$, the parameters are estimated by minimizing the *penalized residual sum of squares (pRSS)* criterion

$$\min_{\tilde{\beta}} \left\{ \sum_{i=1}^{n} (y_i - \tilde{\beta}_0 - \sum_{j=1}^{p} \sum_{\nu=1}^{K_j} \tilde{\beta}_{j\nu} \tilde{\phi}_{j\nu}(x_{ij}))^2 + \tau(\{\lambda^x\}) \right\} \quad , \qquad (3.1)$$

where $\tilde{\beta}_{j\nu}, j = 1, \ldots, p, \nu = 1, \ldots, K_j$, are unknown coefficients and K_j is the number of B-splines for the jth covariate.

Remark 6. The tilde symbol used means that the current expressions are only intermediate steps. In general these expressions cannot be applied in practical use, because e.g. identification conditions (see later in this section) are not integrated.

The expression

$$\tau(\{\lambda^x\}) = \sum_{j=1}^{p} \sum_{\nu=k+1}^{K_j} \lambda_{j\nu} (\Delta^k \tilde{\beta}_{j\nu})^2 \qquad (3.2)$$

is the penalization term. The penaliztion term was first introduced by Eilers & Marx (1996). They suggested to penalize the difference of adjacent coefficients. Hence in (3.1) the expression $\Delta^k \tilde{\beta}_{j\nu}, k = 1, 2, \ldots$, denotes the kth difference, e.g. the 2th difference has the form

$$\begin{aligned} \Delta^2 \tilde{\beta}_{j\nu} &= \Delta^1(\tilde{\beta}_{j\nu} - \tilde{\beta}_{j\nu-1}) \\ &= (\tilde{\beta}_{j\nu} - \tilde{\beta}_{j\nu-1}) - (\tilde{\beta}_{j\nu-1} - \tilde{\beta}_{j\nu-2}) \\ &= (\tilde{\beta}_{j\nu} - 2\tilde{\beta}_{j\nu-1} + \tilde{\beta}_{j\nu-2}). \end{aligned}$$

The parameters $\lambda_{j\nu} \geq 0, \nu = k+1, \ldots, K_j$, with $k = 1, 2, \ldots$, are *local* smoothing parameters that control the amount of shrinkage: the larger the values of $\lambda_{j\nu}$, the larger the amount of shrinkage (Hastie, Tibshirani & Friedman (2001)). If $\lambda_{j,k+1} = \ldots = \lambda_{j,K_j} = \lambda_j$ we have a *global* smoothing parameter for the jth covariate. Although global parameters are more easily to handle, it has been demonstrated by Ruppert & Carroll (2000), that local smoothing parameters yield better performance.

Writing (3.1) in matrix form we obtain

$$pRSS(\boldsymbol{\Lambda}) = (\mathbf{y} - \tilde{\mathbf{B}}\tilde{\boldsymbol{\beta}})^T(\mathbf{y} - \tilde{\mathbf{B}}\tilde{\boldsymbol{\beta}}) + \tilde{\boldsymbol{\beta}}^T\tilde{\mathbf{D}}^T\boldsymbol{\Lambda}\tilde{\mathbf{D}}\tilde{\boldsymbol{\beta}}. \tag{3.3}$$

Here $\tilde{\mathbf{B}}$ is a $n \times (K_1 + \ldots + K_p) + 1$-design matrix

$$\tilde{\mathbf{B}} = [1, \tilde{\phi}_1, \tilde{\phi}_2, \ldots, \tilde{\phi}_p]$$

$$= \begin{bmatrix} 1 & \tilde{\phi}_{11}(x_{11}) & \cdots & \tilde{\phi}_{1K_1}(x_{11}) & \tilde{\phi}_{21}(x_{12}) & \cdots & \tilde{\phi}_{pK_p}(x_{1p}) \\ \vdots & \vdots & & \vdots & \vdots & & \vdots \\ 1 & \tilde{\phi}_{11}(x_{n1}) & \cdots & \tilde{\phi}_{1K_1}(x_{n1}) & \tilde{\phi}_{21}(x_{n2}) & \cdots & \tilde{\phi}_{pK_p}(x_{np}) \end{bmatrix}$$

and $\tilde{\mathbf{D}} = diag(\mathbf{0}, \tilde{\mathbf{D}}_1, \ldots, \tilde{\mathbf{D}}_p)$ is a $[(K_1 - k) + \ldots + (K_p - k)] + 1 \times [K_1 + \ldots + K_p] + 1$-penalization matrix of differences order k, where each matrix $\tilde{\mathbf{D}}_j$ is of dimension $(K_j - k) \times K_j$. In the cases we have differences of first ($k = 1$, left matrix below) or second ($k = 2$, right matrix below) order the matrices $\tilde{\mathbf{D}}_j$ have the structure

$$\tilde{\mathbf{D}}_j = \begin{bmatrix} -1 & 1 & 0 & \cdots & 0 \\ 0 & -1 & 1 & \cdots & 0 \\ \vdots & & \ddots & \ddots & \vdots \\ 0 & \ldots & \ldots & -1 & 1 \end{bmatrix} \quad \tilde{\mathbf{D}}_j = \begin{bmatrix} 1 & -2 & 1 & 0 & \cdots & 0 \\ 0 & 1 & -2 & 1 & \cdots & 0 \\ \vdots & & \ddots & \ddots & \ddots & \vdots \\ 0 & \ldots & \ldots & 1 & -2 & 1 \end{bmatrix}. \tag{3.4}$$

$\boldsymbol{\Lambda} = diag(0, \lambda_{1,k+1}, \ldots, \lambda_{1,K_1}, \lambda_{2,k+1}, \ldots, \lambda_{p,K_p})$ is a smoothing matrix of dimension $[(K_1 - k) + \ldots + (K_p - k)] + 1 \times [(K_1 - k) + \ldots + (K_p - k)] + 1$.

Calculation of the first derivative and setting the expression to zero yields an estimator for $\tilde{\boldsymbol{\beta}}$

$$\hat{\tilde{\boldsymbol{\beta}}}(\boldsymbol{\Lambda}) = (\tilde{\mathbf{B}}^T\tilde{\mathbf{B}} + \tilde{\mathbf{D}}^T\boldsymbol{\Lambda}\tilde{\mathbf{D}})^{-1}\tilde{\mathbf{B}}^T\mathbf{y}. \tag{3.5}$$

and thus an estimator for $E(\mathbf{y})$ is given by

$$\hat{\mathbf{y}} = \tilde{\mathbf{B}}\hat{\tilde{\boldsymbol{\beta}}}(\boldsymbol{\Lambda}). \tag{3.6}$$

The components in an additive model (2.5) are not identifiable without further restrictions. For example let

$$y_i = g(\eta_i) + \epsilon_i$$
$$= f_1(x_{i1}) + f_2(x_{i2}) + \epsilon_i,$$

an additive model with only two covariates. If we add a constant c to the first component f_1 and subtract this constant from f_2 we receive the same additive predictor η_i.

A restriction which yields uniqueness is the expression

$$\sum_{\nu=1}^{K_j} \tilde{\beta}_{j\nu} = 0, \qquad j = 1, \dots, p \tag{3.7}$$

and thus without loss of generality the last coefficient $\tilde{\beta}_{jK_j}, j = 1, \dots, p$, can be represented by a linear combination of the other coefficients, i.e.

$$\begin{aligned}
\tilde{\beta}_{jK_j} &= -\tilde{\beta}_{j1} - \tilde{\beta}_{j2} - \dots - \tilde{\beta}_{j,K_j-1} \\
&= -\sum_{\nu=1}^{K_j-1} \tilde{\beta}_{j\nu}.
\end{aligned} \tag{3.8}$$

With regard to this condition we receive from (3.6) for $y_i, i = 1, \dots, n$,

$$\hat{y}_i = \hat{\tilde{\beta}}_0 + \sum_{\nu=1}^{K_1-1} \hat{\tilde{\beta}}_{1\nu} \tilde{\phi}_{1\nu}(x_{i1}) + \hat{\tilde{\beta}}_{1K_1} \tilde{\phi}_{1K_1}(x_{i1}) + \dots +$$

$$+ \sum_{\nu=1}^{K_p-1} \hat{\tilde{\beta}}_{p\nu} \tilde{\phi}_{p\nu}(x_{ip}) + \hat{\tilde{\beta}}_{pK_p} \tilde{\phi}_{pK_p}(x_{ip})$$

$$\overset{(3.8)}{=} \hat{\tilde{\beta}}_0 + \sum_{\nu=1}^{K_1-1} \hat{\tilde{\beta}}_{1\nu} \tilde{\phi}_{1\nu}(x_{i1}) - \sum_{\nu=1}^{K_1-1} \hat{\tilde{\beta}}_{1\nu} \tilde{\phi}_{1K_1}(x_{i1}) + \dots +$$

$$+ \sum_{\nu=1}^{K_p-1} \hat{\tilde{\beta}}_{p\nu} \tilde{\phi}_{p\nu}(x_{ip}) - \sum_{\nu=1}^{K_p-1} \hat{\tilde{\beta}}_{p\nu} \tilde{\phi}_{pK_p}(x_{ip})$$

$$= \hat{\tilde{\beta}}_0 + \sum_{\nu=1}^{K_1-1} \hat{\tilde{\beta}}_{1\nu} \underbrace{\left(\tilde{\phi}_{1\nu}(x_{i1}) - \tilde{\phi}_{1K_1}(x_{i1}) \right)}_{\equiv \phi_{1\nu}(x_{i1})} + \dots +$$

$$+ \sum_{\nu=1}^{K_p-1} \hat{\tilde{\beta}}_{p\nu} \underbrace{\left(\tilde{\phi}_{p\nu}(x_{ip}) - \tilde{\phi}_{pK_p}(x_{ip}) \right)}_{\equiv \phi_{p\nu}(x_{ip})}.$$

Hence in each matrix $\tilde{\boldsymbol{\phi}}_j = (\tilde{\phi}_{j\nu}), j = 1, \dots, p, \nu = 1, \dots, K_j$, without loss of generality the last (K_j-th) column is subtracted from all the other columns and is not applied in further calculations, anymore.

Writing the last expression in matrix notation we obtain

$$\hat{\mathbf{y}} = \mathbf{B}\hat{\boldsymbol{\beta}}(\boldsymbol{\Lambda}).$$

The design-matrix $\mathbf{B} = [\mathbf{1}, \boldsymbol{\phi}_1, \boldsymbol{\phi}_2, \dots, \boldsymbol{\phi}_p]$ is of dimension $n \times [(K_1 - 1) + \dots + (K_p-1)] + 1$ and $\hat{\boldsymbol{\beta}}(\boldsymbol{\Lambda})$ is a $[(K_1-1) + \dots + (K_p-1)] + 1 \times 1$-coefficient matrix.

Here the estimator $\hat{\boldsymbol{\beta}}(\boldsymbol{\Lambda})$ has the form

$$\hat{\boldsymbol{\beta}}(\boldsymbol{\Lambda}) = (\mathbf{B}^T\mathbf{B} + \mathbf{D}^T\boldsymbol{\Lambda}\mathbf{D})^{-1}\mathbf{B}^T\mathbf{y}, \qquad (3.9)$$

which minimizes the corresponding pRSS criterion

$$pRSS(\boldsymbol{\Lambda}) = (\mathbf{y} - \mathbf{B}\boldsymbol{\beta})^T(\mathbf{y} - \mathbf{B}\boldsymbol{\beta}) + \boldsymbol{\beta}^T\mathbf{D}^T\boldsymbol{\Lambda}\mathbf{D}\boldsymbol{\beta}. \qquad (3.10)$$

Because of the different coefficient vector

$$\hat{\boldsymbol{\beta}}_j = (\hat{\tilde{\beta}}_{j1}, \ldots, \hat{\tilde{\beta}}_{j,K_j-1})^T, j = 1, \ldots, p,$$

it is necessary to adapt the penalization matrix $\tilde{\mathbf{D}}$. The new blockmatrix $\mathbf{D} = diag(\mathbf{0}, \mathbf{D}_1, \ldots, \mathbf{D}_p)$ has dimension $[(K_1 - k) + \ldots + (K_p - k)] + 1 \times [(K_1 - 1) + \ldots + (K_p - 1)] + 1$. The elements $\mathbf{D}_j, j = 1, \ldots, p$, have dimension $(K_j - k) \times (K_j - 1)$ and are computed in the following way:

$$\tilde{\mathbf{D}}_j\hat{\boldsymbol{\beta}}_j = \begin{bmatrix} \tilde{\mathbf{D}}_{j,\,[(K_j-k)-1\times(K_j-1)]} & \vdots & \mathbf{0}_{[(K_j-k)-1\times1]} \\ \hdotsfor{3} \\ \mathbf{0}_{[1\times(K_j-k)-1]} - 1 & \vdots & 1 \end{bmatrix} \cdot \begin{bmatrix} \hat{\tilde{\beta}}_{j1} \\ \vdots \\ \hat{\tilde{\beta}}_{j,K_j-1} \\ -\sum_{\nu=1}^{K_j-1}\hat{\tilde{\beta}}_{j\nu} \end{bmatrix}$$

$$= \begin{bmatrix} \tilde{\mathbf{D}}_{j,\,[(K_j-k)-1\times(K_j-1)]} \cdot \hat{\boldsymbol{\beta}}_j \\ -\hat{\tilde{\beta}}_{j,K_j-1} - \sum_{\nu=1}^{K_j-1}\hat{\tilde{\beta}}_{j\nu} \end{bmatrix}$$

$$= \begin{bmatrix} \tilde{\mathbf{D}}_{j,\,[(K_j-k)-1\times(K_j-1)]} \\ -\mathbf{1}_{[1\times(K_j-2)]} - 2 \end{bmatrix} \cdot \hat{\boldsymbol{\beta}}_j$$

$$= \mathbf{D}_j \cdot \hat{\boldsymbol{\beta}}_j$$

In the cases we have differences of first ($k = 1$) or second ($k = 2$) order the matrices $\mathbf{D}_j$ have the structure

$$\mathbf{D}_j = \begin{bmatrix} -1 & 1 & 0 & \cdots & 0 \\ 0 & -1 & 1 & \cdots & 0 \\ \vdots & \ddots & \ddots & & \vdots \\ 0 & \ldots\ldots & & -1 & 1 \\ -1 & \ldots\ldots & & -1 & -2 \end{bmatrix} \quad \mathbf{D}_j = \begin{bmatrix} 1 & -2 & 1 & 0 & \cdots & 0 \\ 0 & 1 & -2 & 1 & \cdots & 0 \\ \vdots & \ddots & \ddots & \ddots & & \vdots \\ 0 & \ldots\ldots & & 1 & -2 & 1 \\ -1 & \ldots\ldots & & -1 & -1 & -2 \end{bmatrix}.$$

3.2 Penalized Regression Splines with Interactions

For interactions between two metrical variables x_{ir} and x_{is} we have to extend the considerations of the one-dimensional case. As mentioned in (2.9) a two-dimensional function $f_{rs}(x_{ir}, x_{is})$ is represented as a tensor product of two one-dimensional basis functions, i.e.

$$f_{rs}(x_{ir}, x_{is}) = \sum_{\kappa=1}^{K_r} \sum_{\rho=1}^{K_s} \tilde{\gamma}_{rs,\kappa\rho} \tilde{\phi}_{r\kappa}(x_{ir}) \tilde{\phi}_{s\rho}(x_{is}),$$

where K_r and K_s are the numbers of basis functions for the two metrical variables. Similar to the case without interactions the unknown parameters $\tilde{\gamma}_{rs,\kappa\rho}$ are estimated by minimizing the pRSS criterion

$$\min_{\tilde{\gamma}} \left\{ \sum_{i=1}^{n} (y_i - \sum_{r=1}^{p-1} \sum_{s=r+1}^{p} \sum_{\kappa=1}^{K_r} \sum_{\rho=1}^{K_s} \tilde{\gamma}_{rs,\kappa\rho} \tilde{\phi}_{r\kappa}(x_{ir}) \tilde{\phi}_{s\rho}(x_{is}))^2 + \tau(\{\lambda^{xx}\}) \right\},$$

with the penalization term

$$\tau(\lambda^{xx}) = \frac{1}{2} \sum_{r=1}^{p-1} \sum_{s=r+1}^{p} \sum_{\kappa=k+1}^{K_r} \sum_{\rho=1}^{K_s} \lambda_{rs,\kappa\rho}^{(r)} (\Delta^k \tilde{\gamma}_{rs,\kappa\rho})^2 +$$

$$+ \frac{1}{2} \sum_{r=1}^{p-1} \sum_{s=r+1}^{p} \sum_{\kappa=1}^{K_r} \sum_{\rho=k+1}^{K_s} \lambda_{rs,\kappa\rho}^{(s)} (\Delta^k \tilde{\gamma}_{rs,\kappa\rho})^2 \ ,$$

splitting into two parts: the first term yields the penalization of B-splines in r-direction, whereas the second term penalizes B-splines in s-direction. $\lambda_{rs,\kappa\rho}^{(r)}$ and $\lambda_{rs,\kappa\rho}^{(s)}$ are local smoothing paramters for both directions r and s. The indices (r) and (s) only illustrate which direction and covariate is particulary applied and we do not sum up about these indices. If $\lambda_{rs,k+1,\rho}^{(r)} = \ldots = \lambda_{rs,K_r,\rho}^{(r)} = \lambda_{rs,\rho}^{(r)}$ and $\lambda_{rs,\kappa,k+1}^{(s)} = \ldots = \lambda_{rs,\kappa,K_s}^{(s)} = \lambda_{rs,\kappa}^{(s)}$ we have global smoothing parameters for the explanatory variables x_{ir} and x_{is}.

To write the penalization term in matrix form we use matrices of first difference order, as presented in (3.4), in fact

$$\tilde{D}_r = \begin{bmatrix} -1 & 1 & 0 & \cdots & 0 \\ 0 & -1 & 1 & \cdots & 0 \\ \vdots & & \ddots & \ddots & \vdots \\ 0 & \ldots\ldots & & -1 & 1 \end{bmatrix} \qquad \tilde{D}_s = \begin{bmatrix} -1 & 1 & 0 & \cdots & 0 \\ 0 & -1 & 1 & \cdots & 0 \\ \vdots & & \ddots & \ddots & \vdots \\ 0 & \ldots\ldots & & -1 & 1 \end{bmatrix},$$

which have dimensions $(K_r - 1) \times K_r$ for the matrix in r-direction respectively $(K_s - 1) \times K_s$ for the other one. Together with a $j \times j$ identity matrix $\mathbf{I}_j$ we get penalization matrices of difference order $k = 1$ for both directions:

$$\tilde{\mathbf{D}}_{r,1} = \tilde{\mathbf{D}}_r \otimes \mathbf{I}_s \quad \text{with dimension } (K_r - 1)K_s \times K_r K_s,$$
$$\tilde{\mathbf{D}}_{s,1} = \mathbf{I}_r \otimes \tilde{\mathbf{D}}_s \quad \text{with dimension } K_r(K_s - 1) \times K_r K_s.$$

Sometimes we are interested in calculating the penalization matrices of higher difference order. Hence a general notation for differences of order $1 \leq k \leq min\{K_r, K_s\} - 1$ is

$$\tilde{\mathbf{D}}_{r,k} = (\tilde{\mathbf{D}}_{r-k+1} \otimes \mathbf{I}_s) \cdot (\tilde{\mathbf{D}}_{r-k+2} \otimes \mathbf{I}_s) \cdot \ldots \cdot (\tilde{\mathbf{D}}_{r-1} \otimes \mathbf{I}_s) \cdot (\tilde{\mathbf{D}}_r \otimes \mathbf{I}_s)$$
$$= (\tilde{\mathbf{D}}_{r-k+1} \cdot \tilde{\mathbf{D}}_{r-k+2} \cdot \tilde{\mathbf{D}}_{r-1} \cdot \tilde{\mathbf{D}}_r) \otimes \mathbf{I}_s \tag{3.11}$$

for direction r and with dimension $(K_r - k)K_s \times K_r K_s$. A similar expression can be received for direction s

$$\tilde{\mathbf{D}}_{s,k} = (\mathbf{I}_r \otimes \tilde{\mathbf{D}}_{s-k+1}) \cdot (\mathbf{I}_r \otimes \tilde{\mathbf{D}}_{s-k+2}) \cdot \ldots \cdot (\mathbf{I}_r \otimes \tilde{\mathbf{D}}_{s-1}) \cdot (\mathbf{I}_r \otimes \tilde{\mathbf{D}}_s)$$
$$= \mathbf{I}_r \otimes (\tilde{\mathbf{D}}_{s-k+1} \cdot \tilde{\mathbf{D}}_{s-k+2} \cdot \tilde{\mathbf{D}}_{s-1} \cdot \tilde{\mathbf{D}}_s) \tag{3.12}$$

which has dimension $K_r(K_s - k) \times K_r K_s$.

With block matrices $\tilde{\mathbf{D}}_1 = diag(\tilde{\mathbf{D}}_{r,k})$ and $\tilde{\mathbf{D}}_2 = diag(\tilde{\mathbf{D}}_{s,k})$, where $r = 1, \ldots, p - 1, s = r + 1, \ldots, p, r \neq s$, we can write the pRSS criterion in matrix form

$$pRSS(\mathbf{\Lambda}_1, \mathbf{\Lambda}_2) = (\mathbf{y} - \tilde{\mathbf{B}}\tilde{\gamma})^T(\mathbf{y} - \tilde{\mathbf{B}}\tilde{\gamma}) +$$
$$+ \frac{1}{2} \underbrace{\tilde{\gamma}^T \tilde{\mathbf{D}}_1^T \mathbf{\Lambda}_1 \tilde{\mathbf{D}}_1 \tilde{\gamma}}_{\substack{\text{penalization in} \\ \text{first direction}}} + \frac{1}{2} \underbrace{\tilde{\gamma}^T \tilde{\mathbf{D}}_2^T \mathbf{\Lambda}_2 \tilde{\mathbf{D}}_2 \tilde{\gamma}}_{\substack{\text{penalization in} \\ \text{second direction}}} .$$

With calculation of the first derivative the estimator $\hat{\tilde{\gamma}}(\mathbf{\Lambda}_1, \mathbf{\Lambda}_2)$ has the form

$$\hat{\tilde{\gamma}}(\mathbf{\Lambda}_1, \mathbf{\Lambda}_2) = (\tilde{\mathbf{B}}^T \tilde{\mathbf{B}} + \tilde{\mathbf{D}}_1^T \mathbf{\Lambda}_1 \tilde{\mathbf{D}}_1 + \tilde{\mathbf{D}}_2^T \mathbf{\Lambda}_2 \tilde{\mathbf{D}}_2)^{-1} \tilde{\mathbf{B}} \mathbf{y}.$$

Here $\mathbf{\Lambda}_1 = diag(\mathbf{\Lambda}_r)$ and $\mathbf{\Lambda}_2 = diag(\mathbf{\Lambda}_s), r = 1, \ldots, p - 1, s = r + 1, \ldots, p, r \neq s$, are diagonal matrices where each $\mathbf{\Lambda}_r$ has dimension $(K_r - k)K_s \times (K_r - k)K_s$ respectively each $\mathbf{\Lambda}_s$ is of dimension $K_r(K_s - k) \times K_r(K_s - k)$. $\tilde{\mathbf{B}} = diag(\tilde{\phi}_r \otimes \tilde{\phi}_s)$ is a block matrix with tensor products $\tilde{\phi}_r \otimes \tilde{\phi}_s$ which have dimension $n \times K_r K_s$.

Remark 7. Each row of a tensor product represents a point in the grid $\mathbb{R}^2$ which is spanned by the axis of the two directions. Usually there is no observation for each possible grid point. In the following we assume that a tensor product only contains rows for which there are actually observations.

Analogically to the one-dimensional case, also in additive models with interactions there are identification problems by estimating of parameters. To enforce uniqueness of the components, we apply the following restriction to the basis coefficients of each component

$$\sum_{r=1}^{p-1} \sum_{s=r+1}^{p} \sum_{\kappa=1}^{K_r} \sum_{\rho=1}^{K_s} \tilde{\gamma}_{rs,\kappa\rho} = 0. \tag{3.13}$$

Here there are only $K_r K_s - 1$ free coefficients which have to be estimated. The remaining coefficient is unequivocally determined by the other coefficients. Without loss of generality the last coefficient $\tilde{\gamma}_{rs,K_r K_s}$ is calculated by

$$\tilde{\gamma}_{rs,\kappa\rho} = -\sum_{r=1}^{p-1} \sum_{s=r+1}^{p} \left(\sum_{\kappa=1}^{K_r-1} \sum_{\rho=1}^{K_s} \tilde{\gamma}_{rs,\kappa\rho} - \sum_{\rho=1}^{K_s-1} \tilde{\gamma}_{rs,K_r\rho} \right).$$

This restriction has to be considered in the design matrix $\tilde{\mathbf{B}}$ and the penalization matrices $\tilde{\mathbf{D}}_1$ and $\tilde{\mathbf{D}}_2$. In case of $\tilde{\mathbf{B}}$, the last $(K_r K_s$-th) column of the tensor product $\tilde{\phi}_r \otimes \tilde{\phi}_s$ is subtracted from all other columns and is not applied in further calculations, anymore. Thus the resulting matrix $\mathbf{B} = diag(\phi_r \otimes \phi_s)$ has tensor products $\phi_r \otimes \phi_s$ with dimension $n \times (K_r K_s - 1)$.

The modified penalization matrices $\mathbf{D}_{r,1}$ and $\mathbf{D}_{s,1}$ for differences of first $(k = 1)$ order can be generally written as

$$\mathbf{D}_{r,1} = \left[\begin{array}{ccccc} \tilde{\mathbf{D}}_{r-1} \otimes \mathbf{I}_s & \vdots & \mathbf{0}_{[(K_r-2)K_s \times (K_r-1)]} \\ \hdotsfor{5} \\ \mathbf{0}_{[(K_s-1) \times (K_r-2)K_s]} & \vdots -\mathbf{I}_{s-1} & \vdots \mathbf{0}_{[(K_s-1) \times 1]} & \vdots & \mathbf{I}_{s-1} \\ \hdotsfor{5} \\ -\mathbf{1}^T_{[1 \times (K_r-1)K_s]} - \mathbf{e}^T_{[1 \times (K_r-1)K_s]} & \vdots & -\mathbf{1}^T_{[1 \times (K_s-1)]} \end{array} \right],$$

$$\mathbf{D}_{s,1} = \left[\begin{array}{c:c} \mathbf{I}_{r-1} \otimes \tilde{\mathbf{D}}_s & \mathbf{0}_{[(K_r-1)(K_s-1)\times(K_s-1)]} \\ \hdashline \mathbf{0}_{[(K_s-2)\times(K_r-1)K_s]} & \tilde{\mathbf{D}}_{s-1} \\ \hdashline -\mathbf{1}^T_{[1\times(K_r-1)K_s]} & -\mathbf{1}^T_{[1\times(K_s-1)]} - \mathbf{e}^T_{[1\times(K_s-1)]} \end{array}\right].$$

Here $\mathbf{I}_j$ denotes the $j \times j$ identity matrix and $-\mathbf{e}^T_{[1\times j]} = (0,\ldots,0,1)$ is a unit vector of length j with 1 at position $(1,j)$. Using the matrices $\mathbf{D}_{r,1}$ and $\mathbf{D}_{s,1}$ defined above we can also specify penalization matrices for differences of order $1 \leq k \leq min\{K_r, K_s\} - 1$, in fact (compare (3.11) and (3.12))

$$\mathbf{D}_{r,k} = \left[(\tilde{\mathbf{D}}_{r-k+1} \cdot \tilde{\mathbf{D}}_{r-k+2} \cdot \ldots \cdot \tilde{\mathbf{D}}_{r-1}) \otimes \mathbf{I}_s\right] \cdot \mathbf{D}_{r,1}$$

which has dimension $(K_r - k)K_s \times (K_r K_s - 1)$ and

$$\mathbf{D}_{s,k} = \left[\mathbf{I}_r \otimes (\tilde{\mathbf{D}}_{s-k+1} \cdot \tilde{\mathbf{D}}_{s-k+2} \cdot \ldots \cdot \tilde{\mathbf{D}}_{s-1})\right] \cdot \mathbf{D}_{s,1}$$

with dimension $K_r(K_s - k) \times (K_r K_s - 1)$.

Example 1. To illustrate the general formula presented just now we concretely specify the penalization matrices for differences of second $(k = 2)$ order with $K_r = K_s = 4$ basis functions in each direction:

$$\mathbf{D}_{r,2} = \left[\begin{array}{cccccccccccccccc} 1 & & & -2 & & & 1 & & & & & & & & & \\ & 1 & & & -2 & & & 1 & & & & & & & & \\ & & 1 & & & -2 & & & 1 & & & & & & & \\ & & & 1 & & & -2 & & & 1 & & & & & & \\ & & & & 1 & & 0 & -2 & & & 0 & 1 & & & & \\ & & & & & 1 & & 0 & -2 & & & 0 & & 1 & & \\ & & & & & & 1 & 0 & & -2 & 0 & & & & 1 & \\ -1 & -1 & -1 & -1 & -1 & -1 & -1 & 0 & -1 & -1 & -1 & -3 & -1 & -1 & -1 \end{array}\right],$$

$$\mathbf{D}_{s,2} = \left[\begin{array}{cccccccccccccccc} 1 & -2 & 1 & 0 & & & & & & & & & & & & \\ 0 & 1 & -2 & 1 & & & & & & & & & & & & \\ & & & & 1 & -2 & 1 & 0 & & & & & & & & \\ & & & & 0 & 1 & -2 & 1 & & & & & & & & \\ & & & & & & & & 1 & -2 & 1 & 0 & & & & \\ & & & & & & & & 0 & 1 & -2 & 1 & & & & \\ 0 & 0 & 0 & 0 & 0 & 0 & 0 & 0 & 0 & 0 & 0 & 0 & 1 & -2 & 1 \\ -1 & -1 & -1 & -1 & -1 & -1 & -1 & -1 & -1 & -1 & -1 & -1 & -1 & 0 & -3 \end{array}\right].$$

Both matrices have dimension 8×15.

With the modified block matrices $\mathbf{B}$ and $\mathbf{D}_1 = diag(\mathbf{D}_{r,k})$ respectively $\mathbf{D}_2 = diag(\mathbf{D}_{s,k})$, where $r = 1, \ldots, p-1, s = r+1, \ldots, p, r \neq s$, the pRSS criterion has the matrix form

$$pRSS(\mathbf{\Lambda}_1, \mathbf{\Lambda}_2) = (\mathbf{y} - \mathbf{B}\boldsymbol{\gamma})^T(\mathbf{y} - \mathbf{B}\boldsymbol{\gamma}) +$$
$$+ \frac{1}{2} \underbrace{\boldsymbol{\gamma}^T \mathbf{D}_1^T \mathbf{\Lambda}_1 \mathbf{D}_1 \boldsymbol{\gamma}}_{\substack{\text{penalization in} \\ \text{first direction}}} + \frac{1}{2} \underbrace{\boldsymbol{\gamma}^T \mathbf{D}_2^T \mathbf{\Lambda}_2 \mathbf{D}_2 \boldsymbol{\gamma}}_{\substack{\text{penalization in} \\ \text{second direction}}} \, ,$$

which has to be minimized. The result is an estimator

$$\hat{\boldsymbol{\gamma}}(\mathbf{\Lambda}_1, \mathbf{\Lambda}_2) = (\mathbf{B}^T\mathbf{B} + \mathbf{D}_1^T\mathbf{\Lambda}_1\mathbf{D}_1 + \mathbf{D}_2^T\mathbf{\Lambda}_2\mathbf{D}_2)^{-1}\mathbf{B}\mathbf{y}.$$

As mentioned above $\mathbf{B}$ is a block matrix with tensor product matrices $\boldsymbol{\phi}_r \otimes \boldsymbol{\phi}_s$, now of dimension $n \times (K_r K_s - 1)$.

3.3 Estimation of Categorical Covariates within Additive Models

As mentioned in chapter 2.1 the additive model in (2.5) is not restricted to metrical variables. The additive model often includes categorical variables as well as interactions between two categorical variables respectively their interactions with metrical variables. In this section we extend the additive model to terms depending from categorical variables and describe how parameters of categorical terms have to be estimated and integrated in the concept of estimation and penalization.

Suppose we have a dataset consisting of p metrical variables $\mathbf{x}_i = (x_{i1}, \ldots, x_{ip})$ and q categorical variables $\mathbf{z}_i = (z_{i1}, \ldots, z_{iq})$. Each categorical variable z_{ik} can take c_k different categories, i.e. $z_{ik} \in \{1, 2, \ldots, c_k\}$. In the special case of a dichotomous variable there are two different categories, i.e. $z_{ik} \in \{1, 2\}$.

The expected value that a variable z_{ik} takes a certain category is different and can be obtained by using a special coding. We use the 0-1- or *Dummy*-coding. If we assume that the covariate z_{ik} has c_k categories, we can choose one category as reference category. This category can be obtained implicitly by the other $c_k - 1$ categories and is not explicitly specified. Each category can be chosen as reference category. Thus for Dummy-coding with e.g. reference category c_k we have

$$z_{ik}^{(\tau)} = \begin{cases} 1 & \text{if there is category } \tau \text{ of variable } k \\ 0 & \text{else} \end{cases} \qquad \tau = 1, \ldots, c_k - 1.$$

The interpretation of the parameters of a model with $c_k - 1$ Dummy variables is easy to understand if we choose the simpler model

$$E(y_i | z_{ik}) = \beta_0 + \sum_{\tau=1}^{c_k-1} \alpha_k^{(\tau)} z_{ik}^{(\tau)}$$
$$= \beta_0 + \alpha_k^{(1)} z_{ik}^{(1)} + \ldots + \alpha_k^{(c_k-1)} z_{ik}^{(c_k-1)},$$

i.e. a model with only one covariate and without interactions. Then we have parameter values

$$\beta_0 = E(y_i | z_{ik}^{(\tau)} = z_{ik}^{(c_k)}),$$
$$\alpha_k^{(\tau)} = E(y_i | z_{ik}^{(\tau)} = z_{ik}^{(l)}) - E(y_i | z_{ik}^{(\tau)} = z_{ik}^{(c_k)}), \quad l = 1, \ldots, c_k - 1,$$
$$\alpha_k^{(c_k)} = 0,$$

where the last category c_k is chosen as reference category and $\alpha_k^{(c_k)}$ implicitly takes the value zero. The intercept β_0 yields the expectation of the reference category. $\alpha_k^{(\tau)}, \tau = 1, \ldots, c_k - 1$, shows the variation of the expectation of y_i if we change from the reference category $\tau = c_k$ to category $\tau = l, l = 1, \ldots, c_k - 1$. Thus we compare a specific category (i.e. the reference category) with $c_k - 1$ other categories. An example in the field of medicine could be the comparison between a conventional therapy and $c_k - 1$ alternatives.

Hence the additive model of (2.5), which includes terms depending from categorical variables has the form

$$y_i = \beta_0 + \sum_{j=1}^{p} f_j(x_{ij}) + \mathbf{z}_i^T \boldsymbol{\alpha}_i + \sum_{r=1}^{p-1} \sum_{s=r+1}^{p} f_{rs}(x_{ir}, x_{is}) + \sum_{k=1}^{q} z_{ik} \sum_{j=1}^{p} f_j(x_{ij}) + \epsilon_i. \tag{3.14}$$

The term $\mathbf{z}_i^T \boldsymbol{\alpha}_i$ contains the categorical variables and the interactions between two categorical variables. Under the assumption that the last category c_k is the reference category we can write

$$\mathbf{z}_i^T \boldsymbol{\alpha}_i = \underbrace{\sum_{k=1}^{q} \sum_{\tau=1}^{c_k-1} \alpha_k^{(\tau)} z_{ik}^{(\tau)}}_{\text{main effects}} + \underbrace{\sum_{u=1}^{q-1} \sum_{v=u+1}^{q} \sum_{\tau=1}^{(c_u-1)(c_v-1)} \alpha_{uv}^{(\tau)} z_{iu} z_{iv}^{(\tau)}}_{\substack{\text{interactions between two} \\ \text{categorical variables}}}$$

$$= \alpha_1^{(1)} z_{i1}^{(1)} + \ldots + \alpha_1^{(c_1-1)} z_{i1}^{(c_1-1)} + \ldots + \alpha_q^{(c_q-1)} z_{iq}^{(c_q-1)} +$$

$$+ \alpha_{12}^{(1)} z_{i1} z_{i2}^{(1)} + \ldots + \alpha_{12}^{(c_1-1)(c_2-1)} z_{i1} z_{i2}^{(c_1-1)(c_2-1)} + \ldots$$

$$\ldots + \alpha_{q-1,q}^{(c_{q-1}-1)(c_q-1)} z_{i,q-1} z_{iq}^{(c_{q-1}-1)(c_q-1)} \quad .$$

Categorical variables are not developed in basis functions and thus they have no penalization term. For estimation of categorical variables we need a design matrix $\mathbf{B} = (\mathbf{B}^z, \mathbf{B}^{zz})$, where $\mathbf{B}^z$ and $\mathbf{B}^{zz}$ are the parts of the main effects respectively the interactions between different categorical variables. Here $\mathbf{B}$ has the form

$$\mathbf{B} = \begin{bmatrix} z_{11}^{(1)} & \cdots & z_{11}^{(c_1-1)} & \cdots & z_{1q}^{(c_q-1)} & z_{11} z_{12}^{(1)} & \cdots & z_{1,q-1} z_{1q}^{(c_{q-1}-1)(c_q-1)} \\ \vdots & & \vdots & & \vdots & \vdots & & \vdots \\ z_{n1}^{(1)} & \cdots & z_{n1}^{(c_1-1)} & \cdots & z_{nq}^{(c_q-1)} & z_{n1} z_{12}^{(1)} & \cdots & z_{n,q-1} z_{1q}^{(c_{q-1}-1)(c_q-1)} \end{bmatrix} .$$

It is illustrated below how the matrices $\mathbf{B}^z$ and $\mathbf{B}^{zz}$ are integrated in the penalization concept.

The expression $\sum_{k=1}^{q} z_{ik} \sum_{j=1}^{p} f_j(x_{ij})$ of formula (3.14) extends the additive model by multiplicative interaction terms between regressors $z_{ik}, k = 1 \ldots, q$, and metrical covariables $x_{ij}, j = 1, \ldots, p$. The variables x_{ij} (called effect modifiers) change the effects of the regressors z_{ik} via the unspecified function $f_j, j = 1, \ldots, p$. This concept first introduced by Hastie & Tibshirani (1993) is denoted as *varying coefficient model*.

The varying coefficient model can be easily integrated in the penalization concept because it differs from the expression $\sum_{j=1}^{p} f_j(x_{ij})$ by multiplication of a real number, only. Hence the considerations of section 3.1 can be used and adapted to this case.

If we have to estimate an additive model, containing metrical and categorical variables respectively diverse interactions, we have to combine the single parts about estimating in additive models described in the chapter 3. Then the pRSS criterion can be written as

$$pRSS(\mathbf{\Lambda}) = (\mathbf{y} - \mathbf{A}\mathbf{w})^T(\mathbf{y} - \mathbf{A}\mathbf{w}) + \mathbf{w}^T\mathbf{P}^T\mathbf{\Lambda}\mathbf{P}\mathbf{w}. \qquad (3.15)$$

Calculating the first derivative and setting the expression to zero yields an estimator for $\mathbf{w}(\mathbf{\Lambda})$

$$\hat{\mathbf{w}}(\mathbf{\Lambda}) = (\mathbf{A}^T\mathbf{A} + \mathbf{P}^T\mathbf{\Lambda}\mathbf{P})^{-1}\mathbf{A}^T\mathbf{y},$$

and hence an estimator for $\mathbf{y}$

$$\hat{\mathbf{y}} = \mathbf{A}\hat{\mathbf{w}}(\mathbf{\Lambda}).$$

Now we have a look at the form and dimensions of the matrices used in (3.15). The design matrix $\mathbf{A}$ has the form of a block matrix

$$\mathbf{A} = \begin{bmatrix} \begin{matrix} 1 \\ \vdots \\ 1 \end{matrix} \boxed{\mathbf{B}^x} & & & & \\ & \mathbf{B}^z & & & \\ & & \boxed{\mathbf{B}^{xx}} & & \\ & & & \mathbf{B}^{zz} & \\ & & & & \boxed{\mathbf{B}^{xz}} \end{bmatrix}, \qquad (3.16)$$

where the single matrices have the dimensions

- $\mathbf{B}^x \; : n \times [(K_1 - 1) + \ldots + (K_p - 1)]$,

- $\mathbf{B}^z \; : n \times [(c_1 - 1) + \ldots + (c_q - 1)]$,

- $\mathbf{B}^{xx} : n \times [(K_1 K_2 - 1) + \ldots + (K_{p-1} K_p - 1)]$,

- $\mathbf{B}^{zz} : n \times [(c_1 - 1)(c_2 - 1) + \ldots + (c_{q-1} - 1)(c_q - 1)]$,

- $\mathbf{B}^{xz} : n \times [(c_1 - 1) + \ldots + (c_k - 1)] \cdot [(K_1 - 1) + \ldots + (K_p - 1)]$,

where $\mathbf{B}^x$ and $\mathbf{B}^z$ are the design matrices for metrical and categorical variables. $\mathbf{B}^{xx}, \mathbf{B}^{zz}$ and $\mathbf{B}^{xz}$ yield the respective interaction terms. The matrices $\mathbf{B}^z$ and $\mathbf{B}^{zz}$ only contain values 0 or 1.

Remark 8. In formula (3.14) the term $\mathbf{z}_i^T \boldsymbol{\alpha}_i$ includes main effects and interactions of categorical variables. In matrix $\mathbf{A}$ we have (without any loss of generality) divided the two parts for better illustration. Now the two parts are denoted as $\mathbf{B}^z$ and $\mathbf{B}^{zz}$.

The vector $\hat{\mathbf{w}}$ contains the estimators of weights for the single terms of (3.14), i.e. $\hat{\mathbf{w}} = (\hat{\beta}_0, \hat{\boldsymbol{\beta}}, \hat{\boldsymbol{\alpha}}, \hat{\boldsymbol{\gamma}}, \hat{\boldsymbol{\delta}})^T$. The penalization matrix $\mathbf{P}$ is a block matrix

$$\mathbf{P} =$$ 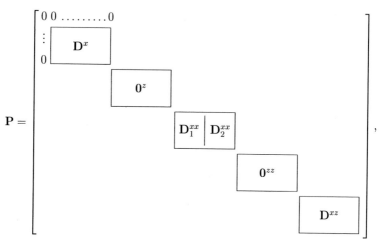 $$,$$

where $\mathbf{0}^z$ and $\mathbf{0}^{zz}$ are zero matrices for the categorical variables (because they have no penalization terms). The dimensions of the single matrices of $\mathbf{P}$ are given by

- $\mathbf{D}^x$: $[(K_1 - k) + \ldots + (K_p - k)] \times [(K_1 - 1) + \ldots + (K_p - 1)]$,
- $\mathbf{0}^z$: $[(c_1 - 1) + \ldots + (c_q - 1)] \times [(c_1 - 1) + \ldots + (c_q - 1)]$,
- $\mathbf{D}_1^{xx}$: $[(K_1-k)K_2 + \ldots + (K_{p-1}-k)K_p] \times [(K_1 K_2 - 1) + \ldots + (K_{p-1} K_p - 1)]$,
- $\mathbf{D}_2^{xx}$: $[K_1(K_2-k) + \ldots + K_{p-1}(K_p-k)] \times [(K_1 K_2 - 1) + \ldots + (K_{p-1} K_p - 1)]$,
- $\mathbf{0}^{zz}$: $([(c_1 - 1)(c_2 - 1) + \ldots + (c_{q-1} - 1)(c_q - 1)])^2$,
- $\mathbf{D}^{xz}$: $[(c_1 - 1) + \ldots + (c_k - 1)] \cdot [(K_1 - k) + \ldots + (K_p - k)] \times$
$\times [(c_1 - 1) + \ldots + (c_k - 1)] \cdot [(K_1 - 1) + \ldots + (K_p - 1)]$.

Finally $\boldsymbol{\Lambda} = diag(0, \boldsymbol{\Lambda}^x, \mathbf{0}^z, \boldsymbol{\Lambda}^{xx}, \mathbf{0}^{zz}, \boldsymbol{\Lambda}^{xz})$ is a block matrix, where (similar to the penalization matrix $\mathbf{P}$) $\boldsymbol{\Lambda}^{xx}$ is splitted into two matrices $\boldsymbol{\Lambda}_1^{xx}$ and $\boldsymbol{\Lambda}_2^{xx}$.

Information Criteria for Model Selection

Information criteria are important tools for model selection. These criteria compare the error of a model with the model complexity (i.e. the number of parameters used). An additional parameter should be integrated in a model only if the value of the information criterion decreases. If we have several competing models, we choose that one with the lowest value of the information criterion.

As seen in the last chapter the performance of the penalized estimate strongly depends on the choice of the smoothing parameters $\lambda_{j\nu}, j = 1, \ldots, p, \nu = 1, \ldots, K_j$. The smoothing parameters have to be chosen in such a way that the criterion used becomes minimal. A criterion commonly used is the *Akaike information criterion (AIC)* introduced by Akaike (1973). This criterion bases on maximization of the logarithm likelihood function (short *log-likelihood*). For derivation of the AIC we start with a linear model, i.e. for a dataset $(y_i, \mathbf{x}_i), i = 1, \ldots, n$, we have

$$y_i = \beta_0 + \sum_{j=1}^{p} \beta_j x_{ij} + \epsilon_i \ , \tag{4.1}$$

where we assume $\epsilon_i \sim \mathcal{N}(0, \sigma^2)$. The likelihood function for normal distribution is given by

$$
\begin{aligned}
L(\boldsymbol{\theta}) &= \prod_{i=1}^{n} f_\theta(\mathbf{x}_i) \\
&= \prod_{i=1}^{n} \frac{1}{\sqrt{2\pi\sigma^2}} e^{-\frac{1}{2}\frac{(y_i - \mu_i)^2}{2\sigma^2}} \ ,
\end{aligned} \tag{4.2}
$$

where $\boldsymbol{\theta} = (\boldsymbol{\mu}, \sigma^2)$ and $\boldsymbol{\mu} = (\mu_1, \ldots, \mu_n)$ is the vector which yields the parameter values. Because of $\mu_i = E(y_i, \mathbf{x}_i) = \hat{y}_i$, inserting of e.g. $\hat{y}_i =$

$\beta_0 + \sum_{j=1}^{p} \beta_j x_{ij}$ we get a parameter vector which consists of the regression coefficients $\beta_0, \beta_1, \ldots, \beta_p$, and one additional parameter σ^2. Application of the natural logarithm to (4.2) yields the log-likelihood

$$l(\boldsymbol{\theta}) = \sum_{i=1}^{n} log f_\theta(\mathbf{x}_i)$$

$$= -\frac{n}{2} log(2\pi) - n \, log(\sigma) - \sum_{i=1}^{n} \frac{(y_i - \hat{y}_i)^2}{2\sigma^2} \quad . \tag{4.3}$$

As estimator for σ^2 we use the maximum likelihood estimator $\hat{\sigma}^2 = (1/n) \sum_{i=1}^{n} (y_i - \hat{y}_i)^2$ which is obtained from (4.3) by calculating the first derivative. Hence (4.3) simplifies to

$$l(\boldsymbol{\theta}) = -\frac{n}{2} log(2\pi) - \frac{2n}{2} log(\hat{\sigma}) - \frac{n\hat{\sigma}^2}{2\hat{\sigma}^2}$$

$$= -\frac{n}{2} log(2\pi\hat{\sigma}^2) - \frac{n}{2} \quad .$$

The AIC has the form

$$AIC_{(1)} = -2l(\boldsymbol{\theta}) + 2q \quad , \tag{4.4}$$

where $q = p+1$ is the number of parameters in the model, which have to be estimated. As distinction from a comparable notation (see formula (4.5)) we index the formula in (4.4) with (1). In search of an appropriate model the $AIC_{(1)}$ has to be minimized. The criterion in (4.4) contains two conflicting parts: if $AIC_{(1)}$ only contains the first term (i.e. the log-likelihood), the criterion becomes smaller in case of a close estimation of the data (if we have an interpolation of the data $AIC_{(1)}$ has the minimal value zero). Thus we would choose that model which has the smallest residual variance estimate σ^2. But the residual variance estimate decreases each time a new variable is added (Miller (2002)). I.e. in case of a criterion consisting of a log-likelihood term, only, also covariates without any influence to the response variable enter the model.

To prevent this problem the $AIC_{(1)}$ contains a second term. This penalization term penalizes complex models, i.e. models which contain too many parameters. Hence the criterion prefers models which have only a few covariates respectively a small number q of parameters. In a nutshell we can say that the $AIC_{(1)}$ chooses models which well approximate the dataset, but do not contain too many variables (which counteracts the danger of interpolation).

In case that the response variable has a normal distribution the $AIC_{(1)}$ in (4.4) can be simplified:

$$
\begin{aligned}
AIC_{(1)} &= -2l(\boldsymbol{\theta}) + 2q \\
&= -2\left[-\frac{n}{2}\,log(2\pi\hat{\sigma}^2) - \frac{n}{2}\right] + 2q \\
&= n\,log(2\pi\hat{\sigma}^2) + n + 2q \\
&= n\,log(\hat{\sigma}^2) + \underbrace{n\,log(2\pi) + n}_{=constant} + 2q \;,
\end{aligned}
$$

where the constant can be ignored for calculation of the best model (because the whole function is only shifted over the constant value in vertical direction and hence the position of the minimum changes in vertical direction but not in horizontal direction). By minimization of the expression the appropriate model is chosen. Several authors use a similar expression which differs by division of n

$$
AIC_{(2)} = log(\hat{\sigma}^2) + 2\frac{q}{n} \;. \tag{4.5}
$$

Here minimization also yields the same best model as above. In future we use the formula (4.5) as Akaike information criterion and symbolize it by AIC (instead of $AIC_{(2)}$).

More generally if we consider an additive model

$$
y_i = \beta_0 + \sum_{j=1}^{p} f_j(x_{ij}) + \epsilon_i \;,
$$

which has any smooth functions $f_j, j = 1,\ldots,p$, we can calculate the number of parameters in the model by the hat matrix $\mathbf{H}$. If the functions f_j are expanded in basis functions, as in the chapters above, the hat matrix (Fahrmeir & Tutz (2001)) has the form (compare (3.9))

$$
\mathbf{H} = \mathbf{B}(\mathbf{B}^T\mathbf{B} + \mathbf{D}^T\mathbf{\Lambda}\mathbf{D})^{-1}\mathbf{B}^T \;.
$$

By replacing the number q of parameters by the trace of the hat matrix we can write the AIC in the form

$$AIC = log\left[\frac{1}{n}\sum_{i=1}^{n}(y_i - \hat{y}_i)^2\right] + 2 \cdot \frac{tr(\mathbf{H})}{n}, \qquad (4.6)$$

where $q = tr(\mathbf{H})$ is the number of parameters involved in the fit. The equality between the number of parameters q and the trace of the hat matrix $tr(\mathbf{H})$ can be motivated by analogy with the linear regression model where we always have $q = tr(\mathbf{H})$ (compare Hastie & Tibshirani (1990)).

Other common criteria which have a similar form as the AIC in (4.6) are the *Bayesian information criterion (BIC)* (Schwarz (1978))

$$BIC = log\left[\frac{1}{n}\sum_{i=1}^{n}(y_i - \hat{y}_i)^2\right] + log(n) \cdot \frac{tr(\mathbf{H})}{n}, \qquad (4.7)$$

the *generalized cross-validation (GCV)* criterion (Craven & Whaba (1979))

$$GCV = log\left[\frac{1}{n}\sum_{i=1}^{n}(y_i - \hat{y}_i)^2\right] - 2 \cdot log\left(1 - \frac{tr(\mathbf{H})}{n}\right), \qquad (4.8)$$

or the T criterion by Rice (1984)

$$T = log\left[\frac{1}{n}\sum_{i=1}^{n}(y_i - \hat{y}_i)^2\right] - log\left(1 - \frac{2 \cdot tr(\mathbf{H})}{n}\right). \qquad (4.9)$$

Another criterion with interesting properties has been proposed by Hurvich & Simonoff (1998) and is called *improved Akaike information (AIC_{imp}) criterion*. It is given by the formula

$$AIC_{imp} = log\left[\frac{1}{n}\sum_{i=1}^{n}(y_i - \hat{y}_i)^2\right] + 2 \cdot \frac{[tr(\mathbf{H}) + 1]}{n - tr(\mathbf{H}) - 2}. \qquad (4.10)$$

To understand the effect of the diverse information criteria we plot the penalization term (i.e. the second term) of each criterion as a function of $tr(\mathbf{H})/n$ (Figure 4.1). Here each solid curve uses $n = 100$ observations. We have to divide between two extreme cases (let the number of observations unchanged):

(1) in case of small values for $tr(\mathbf{H})/n$ (that is equivalent to a small model complexity respectively a large smoothing parameter) the penalty function of all information criteria are comparable. Only the BIC shows a significantly stronger penalization also for very small values of $tr(\mathbf{H})/n$.

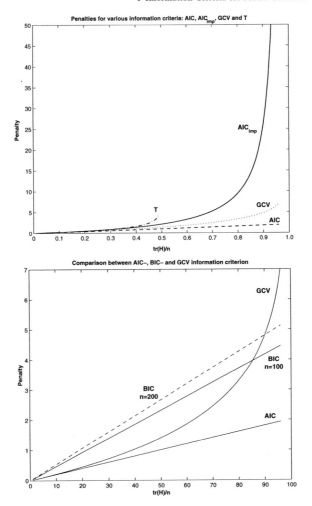

Figure 4.1. *Here penalization term of the information crteria AIC, BIC, GCV, T and AIC_{imp} as function of $tr(\mathbf{H})/n$ is plotted. All solid curves are plotted for the case $n = 100$. The dashed curve shows the BIC with $n = 200$ observations.*

(2) large values for $tr(\mathbf{H})/n$ (i.e. large model complexity respectively a small smoothing parameter) lead to different curve shapes for the diverse criteria. AIC, BIC and GCV show relatively weak penalties and thus there is a tendency of overfitting. In contrast T has a strong

penalty and is closely connected with the danger of underfitting. The improved AIC occupies a position between these two extremes.

In context with variable selection (chapter 8) we often have the problem that a dataset contains redundant or irrelevant variables. The objective is to eliminate these variables from datasets. Thus starting with a large set of variables the number of variables increases during the elimination process. Also the term $tr(\mathbf{H})/n$ takes smaller values in the course of time. Because of the information criteria AIC, improved AIC, GCV, and T have only small penalization for small values of $tr(\mathbf{H})/n$, they often prefer more complex models. Hence the BIC should lead to more accurate results.

Remark 9. Some information criteria (e.g. BIC, improved AIC) depend on the number of observations n. All solid curves in the panels of Figure 4.1 use $n = 100$. In case of an increasing number of observations the curves change their shape or slope. For example the dashed curve in the panel below of Figure 4.1 shows the BIC with a doubled number of observations (i.e. $n = 200$). We realize a larger slope compared with the respective curve for $n = 100$.

In the course of this thesis we compare the different quality of the information criteria by diverse simulation studies.

Part II

Genetic Algorithms for Smoothing
Parameter Choice

5

Genetic Algorithms as a Tool for Optimization

Optimization is essential for almost every field in our daily life. For example in economy we are often interested in minimizing time and costs or maximizing efficacy and utilisation (e.g. of a firm or a hospital). With reference to data collection optimization can be described as a process at which any interesting parameter (e.g. time or efficacy) will be adjusted in a way that we receive a minimal or maximal result. The process or function is known as cost function, objective function or fitness function whereas the output is the cost or fitness.

In context with the choice of smoothing parameters there is also an important optimization problem. For an additive model without interactions (compare section 2.3), p global smoothing parameters have to be chosen. For local smoothing the number of parameters increases to $K_1 + \ldots + K_p$ which results in $30p$ smoothing parameters for 30 knots in each dimension. Thus there is a complex problem to optimize all smoothing parameters.

There are several very efficient algorithms which can be used to solve any optimization problem. They can be subdivided into two principal categories:

(1) Algorithms using the gradient of an error function (this can be e.g. the mean squared error function) for optimization. All these methods start with an initial estimate for the parameter ω and design a sequence of approximations $\omega_{(2)}, \omega_{(3)}, \ldots, \omega_{(t)}$. The goal is that $\omega_{(t)}$ converge against the optimal parameter value ω^*, i.e. $\omega_{(t)} \longrightarrow \omega^*$ as $t \longrightarrow \infty$. Thereby we always change the parameters $\omega_{(t)}$ with regard to a decrease of the error function. Examples for this type of algorithm are gradient decent algorithm, Newton's method and Quasi-Newton's

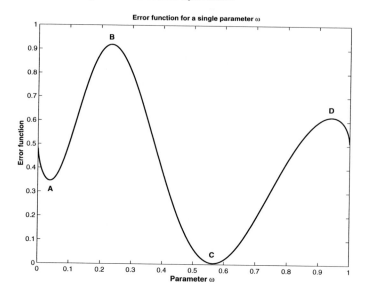

Figure 5.1. *Error function for a single parameter ω. We are interested in finding the global maximum at point B. In this example point A and point C are local minima, whereas point D is a local maximum.*

method, line search, backpropagation and the Levenberg-Marquardt algorithm (Bishop (1995), Haupt & Haupt (1998)).

(2) Algorithms which do not use the gradient of an error function. Examples of these so called combinatorial optimization methods are evolutionary algorithms (see later) or simulated annealing (Kirkpatrik, Gelatt, Jr. & Vecchi (1983)).

For many optimization problems the methods of the first group (in future, we call them classical methods) are excellent and quick-solution-procedures. For instance, they are superior to the combinatorial methods for well-behaved convex analytical functions of only a few parameters. In reality, however, many realistic problems do not fall into this category. This is because we often have to optimize many parameters simultaneously and most optimization problems have several local optima apart from the global optimum (Figure 5.1).

As classical methods select their parameter values with regard to a decrease of the error function we have the danger of a premature convergence in a local optimum.

In search of new strategies of optimization a variety of methods has been suggested in the 1960's, each of them based on Darwin's evolution theory (Darwin (1859)). These methods are known as *evolutionary algorithms* and refer to the principle that better adapted (more fit) individuals win against their competitors under equal conditions. Like their biological standard evolutionary algorithms use biological components (or operators) like selection, crossover, or mutation to model the natural phenomenon of genetic inheritance and Darwinion strife of survival.

The following advantages explain why evolutionary algorithms are more efficient to solve complex optimization problems in comparison with classical methods: evolutionary algorithms

- do not require knowledge or gradient information about the response surface.

- perform very well for complex large-scale optimization problems. They can jump out of a local optimum.

- deal with a large number of parameters.

- have less difficulties with response surfaces which have discontinuities.

- search the search space at several locations simultaneously and hence provide a list of potential parameters. In each other method the solution is only one point in the search space.

5.1 The Language of Evolutionary Algorithms

For further understanding we give a short introduction into the language of evolutionary algorithms. These algorithms use a vocabulary borrowed from natural genetics. Some background of the biological processes of genetics and the origin of terminology is provided by Haupt & Haupt (1998), and Mitchell (1996).

The function to be optimized is denoted as fitness function. The optimization problem can be treated as a minimization- or a maximization

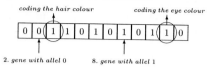

Figure 5.2. *The string consists of twelve genes (loci = 1,...,12). Each gene can obtain the two alleles 0 or 1.*

problem. Without any loss of generality we assume maximization problems only, because minimizing a function f is equivalent to maximizing a function $-f$, i.e.

$$min\ f(x) = max[-f(x)].$$

The smallest units linked with relevant information of a genetic algorithm are called *genes* (or *features, characters, decoders*). Genes are either single units or short blocks of adjacent units and the information is coded in the shape of numbers, characters, or other symbols. With real-coded evolutionary algorithms every gene is a single unit coded by a real value. Usually several genes are arranged in linear succession called *string* (also *chromosome, individual*). In the context of smoothing parameter selection for additive models as in (2.3) a string is a vector of the form $(\lambda_{1,k+1},\ldots,\lambda_{1,K_1},\lambda_{2,k+1},\ldots,\lambda_{p,K_p})$. Here we have considered the penalization as described section 3.1. Thus a local smoothing parameter $\lambda_{j\nu}$ corresponds to one gene. In the case of global smoothing parameters a string reduces to $(\lambda_1,\ldots,\lambda_p)$.

Genes often occur in several distinct states which we call *alleles* (*feature values*). For instance one simple kind of evolutionary algorithm takes alleles 0 and 1; for larger alphabets more alleles are possible at each fixed position (*locus*) of a string. The combination of alleles of a string is known as *genotype*. By application of operators (e.g. crossover and mutation, described in section 5.2) to a string, some alleles are changed and thus the string has a modified genotype. A modification of the configuration of a genotype usually yields an observable change of an individual's trait. The trait is called *phenotype*. In natural genetics eye colour is an example of a phenotype (see Figure 5.2).

Evolutionary algorithms yield several strings as a potential solution of an optimization problem. This collection of strings is named a *population*. As mentioned above the application of operators like crossover and mutation can generate new strings with a different genotype. This new population of strings is called *offspring*. We denote the particular populations as *generations* or, more precisely, as parent- respectively offspring generation.

Evolutionary algorithms can be subdivided into three principal research lines:

- *Evolution Strategies* (Rechenberg (1973); Schwefel (1975); Schwefel (1995)): classical evolution strategies work with populations consisting of one string only. Each gene of a string is represented as a pair $(\mathbf{x}_t, \boldsymbol{\sigma})$ of real-valued vectors. The first vector $\mathbf{x}_t$ (iteration t) represents a point in the search space and the second vector $\boldsymbol{\sigma}$ is a vector of standard deviations. As genetic operator, "mutation" is only used and realized by

$$\mathbf{x}_{t+1} = \mathbf{x}_t + \mathcal{N}(0, \boldsymbol{\sigma}),$$

  where $\mathcal{N}(0, \boldsymbol{\sigma})$ is a vector of independent normally distributed numbers with a mean of zero and standard deviations $\boldsymbol{\sigma}$. The mutated string replaces its parent in case of better values of the fitness function. Otherwise the offspring is eliminated and the population remains unchanged. In course of time also populations with more than one string and additional operators (e.g. crossover) have been used in evolutionary strategies.

- *Evolutionary Programming* (Fogel, Owens & Walsh (1965)): this evolutionary technique has orginally been used in context with artificial intelligence. The idea was to construct an evolutionary algorithm which predicts an unknown state in a time series on the basis of previously known states. The representations used depend on the given problem (e.g. real-valued vector). In evolutionary programming the only genetic operator is the mutation operator applied to each parent string. By a selection procedure half of the fittest strings of the whole population (parents and offsprings) are selected and form the next generation. It should be noticed that an evolutionary programming with real-value representation is similar to evolutionary strategies without crossover.

- *Genetic Algorithms* (Holland (1975); Goldberg (1989)): this technique is described in detail later.

While the particular research directions were distinct at the beginning, they started to mix later, and it became sometimes difficult to differentiate between various directions. But one differentiation is possible even today:

in contemporary evolutionary programming algorithms crossover is not used and thus mutation is the decisive genetic search operator. In the context of this thesis we restrict ourselves to genetic algorithms.

5.2 The Basic Elements of Genetic Algorithms

5.2.1 Binary and Real-coded Representations

First one has to find a suitable representation of strings to apply genetic algorithms. Well-known codings are the *floating-point* (also *real-coded*) representation and the *binary* representation.

To descibe the binary representation we suppose that we wish to optimize a function of p variables, i.e.

$$f: \quad \begin{matrix} \mathbb{R}^p & \longrightarrow & \mathbb{R}^1 \\ (x_1, \ldots, x_p) & \mapsto & f(x_1, \ldots, x_p) \end{matrix} \quad , \tag{5.1}$$

where each variable x_i can take values from a domain $D_i = [a_i, b_i] \subseteq \mathbb{R}^1$ and $f(x_i, \ldots, x_p) \geq 0$ for all $x_i \in D_i$. Here we consider the case that the variables' values of function f have a precision of e.g. six decimal places. To achieve this precision we must divide each domain D_i into $(b_i - a_i) \cdot 10^6$ equally sized ranges. Here

$$(b_i - a_i) \cdot 10^6 \leq 2^{m_i} - 1 \tag{5.2}$$

fulfils the precision requirement where m_i is the smallest integer which fits the inequation. Hence every value x_i is presented by a string-segment of m_i genes and the whole string consists of $m = \sum_{i=1}^{p} m_i$ genes. To calculate the real value x_i of such a string-segment we choose the formula

$$x_i = a_i + \frac{(b_i - a_i)}{2^{m_i} - 1} \cdot \left(\sum_{j=0}^{m_i - 1} g_i^j \cdot 2^j \right) \quad , \tag{5.3}$$

where g_i^j denotes the value of gene i at position $j = 0, 1, \ldots, m_i - 1$.

Example 2. In Figure 5.3 the function $f(x_1, x_2)$ has a precision of three decimal places for the variables' values. If for example x_1 has a domain $D_1 = [0, 1]$ the inequality (5.2) yields $1000 = (1 - 0) \cdot 10^3 \leq 2^{10} - 1 = 1023$ and thus $m_1 = 10$. With a domain $D_2 = [0, 5]$ for x_2 we compute $5000 =$

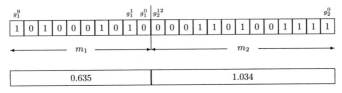

Figure 5.3. *In this Figure the precision of function $f(x_1, x_2)$ with two variables is shown. The string above has binary representation, the other one real-coded representation. Details in the text.*

$(5-0) \cdot 10^3 \leq 2^{13} - 1 = 8191$, i.e. $m_2 = 13$. The associated real value x_1 is computed for the first ten genes, i.e. $x_1 = 0 + \frac{1-0}{2^{10}-1} \cdot (0 \cdot 2^0 + 1 \cdot 2^1 + \cdots + 1 \cdot 2^9) = 0.635$. In analogy we obtain $x_2 = 1.034$ for the next thirteen genes.

Goldberg (1991) shows that binary coded strings with fixed length are the most effective representation. This coding also facilitates theoretical analysis and allows elegant operators. These are the main reasons why binary string representation of solutions has dominated genetic algorithm research over the time.

An interesting alternative for binary coding is the *real-coded* (or *floating-point*) representation. In this case a string consists of floating numbers and has the same length as the solution string. Hence unlike binary coding each variable is coded by one allele only (see Figure 5.3).

On principle real-coded strings have essential advantages:

- *Saving of time:* The length of real-coded strings is much shorter compared to binary-coded strings. Therefore the number of necessary operations (e.g. mutation, crossover, coding, decoding,...) is by far lower.

- *Space complexity:* The use of real parameters facilitates the use of large (and also unknown) domains for the variables. This is limited for strings with binary representation. Suppose we have an optimization problem with 100 variables which can take values from a domain $[-10, 10]$. In addition we require a precision of six decimal places for the variables' values. With the formula (5.2) we compute a string-length of 2500 alleles (i.e. 25 alleles for each variable) for the solution vector. Hence there is an enormous time complexity for one iteration step. Furthermore we need many iteration steps to generate an adequate and secured solution. For this reason binary representation is only applied in low complex spaces.

- *High precision:* Real-coded representation yields solutions of high precision. The level of precision is only limited by the type of the computer used.

- *Closer to the problem space:* Many properties of the problem space can be utilized in real-coded representation. For example: Two points close to each other in representation space also take this property in the problem space and vice versa. This fact does not necessarily occur in binary representation mainly because the distance is generally described through the number of different alleles. Suppose we have two variables which are close to each other and can take values from a domain $[0, 1]$, e.g. $x_1 = 0$ and $x_2 = 0.12$. The values in binary representation are $(0\,0\,0\,0\,0\,0\,0\,0\,0\,0)$ and $(0\,0\,0\,1\,1\,1\,1\,1\,1\,1)$, respectively.

- It is easier to design operators which incorporate problem specific knowledge.

5.2.2 Structure of a "classical" Genetic Algorithm

Before starting the iterative genetic algorithm the user needs to construct an initial population of several strings. This population usually consists of genes chosen randomly from a uniform distribution on a given interval. The population size (noted as *popsize*) is usually chosen freely. A rating of the quality of the used smoothing paramater combination is given by an information criterion as mentioned in chapter 4. Here, the criterion value has to be chosen in such a way that the criterion becomes minimal. For use of the genetic algorithm it is more suitable to work with a criterion which has to be maximized. This is easily achieved by simple mathematical transformations. For that purpose we subtract a sufficiently large constant from all criterion values of a population in such a way that all the values become negative. Simulations in section 7.1 show that the largeness of the chosen constant has no influence on the results. Following multiplication with (-1) yields a criterion which has to be maximized. We denote the values which characterize the quality of the strings as *fitness values* (short *fitness*).

For the design of powerful genetic algorithms, operators like crossover, mutation or selection are important. The reason is that operators control the generation of offsprings which ideally yields better solutions of our optimization problem. First we give a short description of how the operators work.

- *Crossover*: this operator combines the genes of two randomly chosen parent strings and yields similar children strings. Thus features and information of the parent strings are transmitted to the offsprings.

  Example 3. A simple crossover operator is a *one-point crossover*. Suppose we have selected two strings (each has m alleles) for crossover

$$String\ 1 \quad (b_1 b_2 \ldots b_{pos} | b_{pos+1} \ldots b_m)$$
$$String\ 2 \quad (c_1 c_2 \ldots c_{pos} | c_{pos+1} \ldots c_m)$$

  We choose a random number *pos* of $[0, 1, \ldots, m-1]$ which points out the position of our crossover-point. By swapping corresponding segments both parent strings are replaced by a pair of their offspring:

$$Offspring\ 1 \quad (b_1 b_2 \ldots b_{pos} | c_{pos+1} \ldots c_m)$$
$$Offspring\ 2 \quad (c_1 c_2 \ldots c_{pos} | b_{pos+1} \ldots b_m)$$

- *Mutation*: Here a change of the alleles of randomly chosen genes introduces some extra variability into the population.

  Example 4. In case of a string with binary representation (as seen in Figure 5.2) the allele "0" of the second gene (randomly chosen for mutation) would become "1".

- *Selection*: In the selection process strings with good fitness values have larger probability of remaining in the population. On the other side strings with bad fitness values are removed with large probability. Thus the selection operator tries to improve the fitness of a population by decreasing the strings' diversity.

After executing the steps described above, we get a new population for which we must calculate the quality (i.e. the fitness) of each string again. Then we start a new iteration step. The number of iteration steps is determined by a termination condition. Altogether we can differentiate them into three main classes:

(1) *Heuristic termination conditions*: here the search is terminated after a predefined number T of generations, i.e. if the current generation number $t > T$ the algorithm stops. This termination condition assumes that the user knows about the characteristics of the function which influences the number of generations. But this assumption is given very seldom.

(2) *Genotype termination conditions* are based on the string structure and check the number of converged alleles (an allele is converged if a default number of strings in a population has the same value in this allele). If the percentage of converged alleles exceeds a predefined constant the search will be terminated.

(3) *Phenotype termination conditions* measure the progress (or fitness) of the strings in a population. If the progress is under a lower limit the search will be terminated.

The iteration step ends by selecting new strings from the total of strings forming the population. This selection aims at obtaining a constant number of strings in a population. If the population has a size of μ parent strings which generate λ offsprings the problem arises, which of the $(\mu+\lambda)$ strings will enter the new population. In case we are restricted to the fitness-criterion as the only feature of goodness we need to choose the μ best strings as new parents. There are different possibilities of selection:

- The μ best strings are selected from the offspring ((μ,λ)-*strategy*).

- The μ best strings are selected from the whole population consisting of the new and the old strings ($(\mu+\lambda)$-*strategy* or *elitist*-strategy). Thus the best solutions of the old strings are not forgotten.

The whole process of iteraton is summarized in Figure 5.4.

5.3 Genetic Algorithms with Adaptive Operators

In context with the "classical" genetic algorithm (section 5.2.2) we generally outlined some important operators, in fact crossover, mutation, and selection. Now we give some details how these operators have to be chosen.

It is common practice to use *uniform* operators as a tool in genetic algorithms, i.e. operators with unchangeable shape and influence during

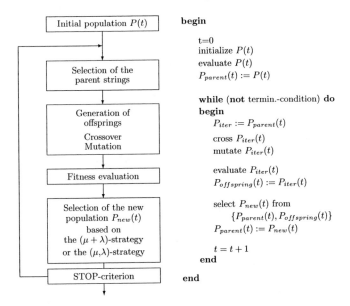

Initial population $P(t)$	**begin**
	t=0
	initialize $P(t)$
	evaluate $P(t)$
Selection of the parent strings	$P_{parent}(t) := P(t)$
	while (**not** termin.-condition) **do**
Generation of offsprings	**begin**
Crossover Mutation	$P_{iter} := P_{parent}(t)$
	cross $P_{iter}(t)$
	mutate $P_{iter}(t)$
Fitness evaluation	evaluate $P_{iter}(t)$
	$P_{offspring}(t) := P_{iter}(t)$
Selection of the new population $P_{new}(t)$ based on the $(\mu + \lambda)$-strategy or the (μ, λ)-strategy	select $P_{new}(t)$ from $\{P_{parent}(t), P_{offspring}(t)\}$ $P_{parent}(t) := P_{new}(t)$ $t = t + 1$ **end**
STOP-criterion	**end**

Figure 5.4. *Structure of a genetic algorithm in summary, given as a flowchart (left side) and as a computer program (right side).*

application of a genetic algorithm. But several authors show that many optimization problems are inadequately solved by unflexible and rigid uniform operators. The reason is that operators have to cope with different tasks (compare the exploitation-exploration-dilemma in section 5.3.1) with various purposes during the application of a genetic algorithm (Herrera, Lozano & Verdegay, 1998; Michalewicz, 1996). To solve these different problems adequately, we require operators which may change during the genetic algorithm. We call these operators *adaptive* (or *non-uniform*) operators.

There are several possibilities to form adaptive operators:

- Change of the default probabilities of selection, crossover and mutation respectively other parameters (e.g. population size) at different times during the genetic algorithm.

- Construction of an operator, which considers the different tasks during application of the genetic algorithm.

Modelling of suitable operators basically depends on the dilemma between the two conflicting objectives exploitation and exploration, which is described in the following section.

5.3.1 Exploitation-Exploration-Dilemma

In each iteration step with a probability > 0 the genetic algorithm intends to produce offsprings which are more fit compared to their parents. The search for suitable offsprings is largely influenced by two conflicting objectives: Exploiting the best (most fit) solutions and exploring the search space. If we want to solve an optimization problem with a genetic algorithm the initial population very seldom includes strings with solutions at (or at least close to) the global optimum. Thus we intend to generate offsprings which are scattered over the whole search space and thereby we hope that at least one of the strings is located near the global optimum. The property to explore the search space with strings and acquire information about the nature of the space is described by the term *exploration*.

After some iteration steps the genetic algorithm may have generated new strings located closer to the global optimum. In this case we are primarily interested in obtaining information near the optimum by utilizing the local possibilities of upgrade close to the parents and by generating more fit offsprings there. This stepwise improvement of local information is called *exploitation*.

Usually previous knowledge about the location of the global optimum (or optima) does not exist. An important fact of all optimization problems is the danger of a premature convergence. If the parameters which characterize a genetic algorithm (e.g. crossover- or mutation probabability) are constant during the iterations the algorithm may be trapped in a local optimum and has little chance to leave it. In these cases we have an undesirable premature convergence with relatively bad solutions. To solve this problem we have to generate strings in other areas of the search space (exploration). In contrast, if we are near to a global optimum we would prefer a convergence (exploitation).

Hence we realize that we need to find a suitable balance between exploration and exploitation during the whole iteration process. The relevance of these two conflicting objectives is thereby differently weighted at various times. For example at the beginning (where we have no idea about the location of the global optimum) of the genetic algorithm exploration is more relevant compared with exploitation and vice versa. In the following section some operators referring to the deliberations of the exploitation-exploration-dilemmas are described.

5.3.2 Improved Arithmetical Crossover (IAC)

In the last decade numerous different types of crossover operators have been suggested (see e.g. Eshelman & Schaffer (1993), Michalewicz (1996), Radcliffe (1991a), Wright (1991)). A survey on several simulations can be found in Herrera, Lozano & Verdegay (1998). Here we present a crossover operator called *improved arithmetical crossover*, shortly *IAC* (Krause & Tutz (2003)).

Suppose we have two real-coded strings (each has m genes) for crossover with values in an interval $[l_{lo}, l_{up}]$ with lower limit l_{lo} and upper limit l_{up}

$$String\ 1\ (a_1 \ldots a_i \ldots a_m)$$
$$String\ 2\ (b_1 \ldots b_i \ldots b_m).$$

The IAC operator is defined by (compare also Figure 5.5)

$$
\begin{aligned}
c_i &= \nu a_i + (1 - \nu) b_i, \\
d_i &= b_i + \delta_1 (l_{up} - b_i), \\
e_i &= a_i - \delta_2 (a_i - l_{lo}),
\end{aligned}
\tag{5.4}
$$

with $i = 1, \ldots m$, and thus the offsprings have the form

$Offspring\ 1\ (\nu a_1 + (1 - \nu) b_1 \ldots \nu a_i + (1 - \nu) b_i \ldots \nu a_m + (1 - \nu) b_m)$

$Offspring\ 2\ (b_1 + \delta_1 (l_{up} - b_1) \ldots b_i + \delta_1 (l_{up} - b_i) \ldots b_m + \delta_1 (l_{up} - b_m))$

$Offspring\ 3\ (a_1 - \delta_2 (a_1 - l_{lo}) \ldots a_i - \delta_2 (a_i - l_{lo}) \ldots a_m - \delta_2 (a_m - l_{lo}))$

where $\nu \in [0, 1]$ can be chosen constantly or variably over the number of iterations. The parameters $\delta_i \in [0, 1], i = 1, 2$, are uniformly distributed random numbers. Every string takes values in the default interval $[l_{lo}, l_{up}]$.

A freely chosen crossover probability p_c determines which strings of the parent population shall be selected for crossover. Therefore we generate a random (float) number $r_i \in [0, 1], i = 1, \ldots, popsize$, for every string of the population. A string is used for crossover operation if $r_i < p_c$ holds. In

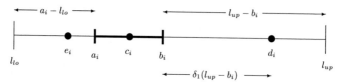

Figure 5.5. *With a_i and b_i representing the parents the IAC operator generates the appropriate genes c_i, d_i and e_i of the children. The first offspring c_i is located within the parents' interval $[a_i, b_i]$. The other children are randomly positioned left and right outside the interval $[a_i, b_i]$. Every string only takes values within the range between l_{lo} and l_{up}.*

the crossover process we need couples of strings and thus it is necessary to select an even number of parent strings.

The IAC operator generates three new offsprings and we select the two best strings to replace the parents. Interestingly, the IAC operator yields children which improve exploration and exploitation simultaneously. Figure 5.5 shows two offsprings (d_i and e_i), located outside the parents' interval $[a_i, b_i]$ and thus regions further apart in the search space can be explored. In addition, one child (here c_i), located within the parents' interval is primarily responsible for an improvement of exploitation.

In section 7 we compare the quality of the IAC operator with that of the arithmetical crossover operator (Michalewicz (1996)). Thereby the arithmetical crossover operator is defined by a weighted linear combination of two parents, i.e.

$$
\begin{aligned}
c_i &= \nu a_i + (1 - \nu)b_i, & i = 1, \ldots, m \\
d_i &= \nu b_i + (1 - \nu)a_i, & i = 1, \ldots, m
\end{aligned}
\tag{5.5}
$$

where $\nu \in [0, 1]$ can be chosen constantly (uniform arithmetical crossover) or variably over the number of iterations (non-uniform arithemical crossover). The arithmetical crossover operator generates two children. Depending on the choice of parameter ν, they only take values in the parents' interval $[a_i, b_i]$. Since the childrens' position is relatively close to their parents, it enhances exploitation. The consequence is missing exploration and thus large parts of the search space remain unconsidered.

5.3.3 Non-uniform Mutation

As mentioned above the purpose of the mutation operator is to introduce some extra variability into the population. Several types of mutation operators have been developed (see e.g. Davis (1991), Michalewicz (1996), Mühlenbein & Schlierkamp-Voosen (1993), Voigt & Anheyer (1994)). A survey on various examples of simulations can be found in Herrera, Lozano & Verdegay (1998), and Michalewicz (1996). In our genetic algorithm we use the *non-uniform mutation operator* presented by Michalewicz (1996).

For every gene of a string we generate a random number $r_{gene} \in [0, 1]$ and compare r_{gene} with a default probability p_m. If $r_{gene} < p_m$, the gene mutates, i.e. it changes its value. Suppose we have a string $(a_1 \ldots a_i \ldots a_m)$ of length m and randomly select the gene a_i for application of the non-uniform mutation operator. Then we get a vector $(a_1 \ldots a_i' \ldots a_m)$ where

$$a_i' = \begin{cases} a_i + (l_{up} - a_i)(1 - r^{(1-\frac{t}{T})^b}) & if \quad \tau = 0 \\ a_i - (a_i - l_{lo})(1 - r^{(1-\frac{t}{T})^b}) & if \quad \tau = 1 \end{cases} . \tag{5.6}$$

Here τ is a random number which may have a value of zero or one, $r \in [0, 1]$ is a uniform random number, T is the maximum number of generations and b is a user-dependent system parameter which determines the degree of non-uniformity. The function

$$g(t) = (1 - r^{(1-\frac{t}{T})^b}) \tag{5.7}$$

yields values in the interval $[0, 1]$ and with (5.6) we get (compare also Figure 5.6)

$$(l_{up} - a_i)g(t) \in [0, l_{up} - a_i] \tag{5.8}$$
$$(a_i - l_{lo})g(t) \in [0, a_i - l_{lo}] .$$

We can distinguish between two extreme cases which are illustrated on Figure 5.7 (without any loss of generality we have restricted our attention to $\tau = 0$):

(1) **t small**: in this case the exponent in (5.7) yields a value close to one and thus $g(t)$ is primarily influenced by a suitable choice of the random number r. Figure 5.7 illustrates this context: for a generation number $t = 5$ we have approximately a straight line with slope -1. As the random number r is uniformly distributed each value $g(t)$ can be (approximately) accepted with the same probability. Hence it also follows

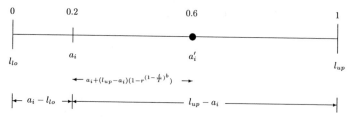

Figure 5.6. *Gene a_i and its mutated offspring a'_i in a default domain $[l_{lo}, l_{up}] = [0, 1]$ are shown here. To illustrate the function $g(t)$ we have chosen $\tau = 0$ without loss of generality. For the parameters $t = 1$, $T = 10$, $b = 1$ and $r = 0.21$ we get $g(t) = 0.75$ and thus $a'_i = 0.6$.*

that every $a'_i \in [0, l_{up} - a_i]$ in (5.6) has nearly the same probability to be taken.

If t increases ($t = 60$ in Figure 5.7) the uniform probability of acceptance for a'_i decreases.

(2) **t large:** if the generation number t is large $g(t)$ in (5.7) (and also the formula in (5.8)) obtains values close to zero for a wide range of random numbers r. This is illustrated in Figure 5.7 for $t = 95$. Thus there is a tendency that the offspring a'_i in (5.6) are close to his parent a_i.

It should also be noted again that there always exists a small probability that

- if $r \to 1$ case (1) can yield offsprings close to a_i;

- if $r \to 0$ case (2) can yield offsprings far away from a_i.

In a nutshell we can say that the genetic algorithm initially explores the whole search space right from the parents' interval uniformly for an accurate a'_i. However at a later stage of the algorithm we primarily prefer those a'_i which are close to their parent a_i.

5.3.4 Sampling Methods

Two important issues exist in genetic search: on the one hand we need several distinguishable strings (that means a large *population diversity*) to search successfully for a global optimum in the search space (exploration). On the other hand a suitable selection of promising strings (we denote this as increase of *selective pressure*) yields a faster convergence of the genetic

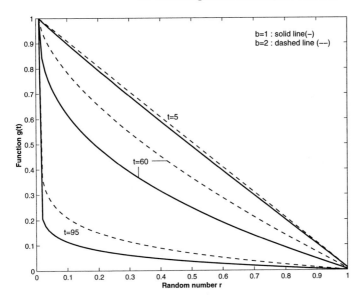

Figure 5.7. *Here the function g(t) in subject to a uniform distributed random number r for three sizes of the generation number t is shown. For each generation number t we have plotted two curves with different parameters b. The larger parameter b, the lower is the degree of non-uniformity.*

search (exploitation). These factors are closely linked because an increase of selective pressure results in a decrease of the diversity of population and vice versa. Hence strong selective pressure supports premature convergence in a local optimum while a weak selective pressure can make the search ineffective. For a suitable balance we need sampling methods which try to select accurate strings of a population at each iteration step of the algorithm.

In literature there are many suggestions of sampling methods. The most famous methods are probably *stochastic universal sampling* (Baker, 1987), *rank-based* techniques (Baker (1985), Whitley (1989)) and *tournament selection* (Goldberg, Deb & Korb (1991)). Here we introduce a new modification of the stochastic universal sampling (Baker (1985)). Our *modified selection procedure* (*modSP*), including crossover- and mutation operators, consists of six steps and is illustrated in Figure 5.8:

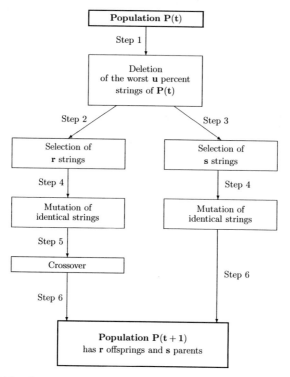

Figure 5.8. *Structure of the modified selection procedure (modSP) given as a flowchart. Details in the text.*

Step 1: Suppose that a population $P(t)$ is generated in iteration step t. Then delete the worst u percent strings of $P(t)$.

Step 2: From the remaining strings of step 1 randomly select r strings, which do not necessarily have to be distinct.

Step 3: From the remaining strings of step 1 randomly select s parent strings. These have not to be distinct from the r selected strings in step 2.

Step 4: If a string has one or more further identical strings in the population (i.e. all genes of the strings are identical) the copies will be mutated. How many genes of a string are randomly selected and mutated is controlled by a random number (at least one gene is mutated). After mutation, there are r respectively s

different strings. This operation will also be executed for the s parent strings.

Step 5: Controlled by the crossover probability p_c, apply a crossover operator to the set of the r (distinct) strings and generate $i, 2 \leq i \leq r$ new strings.

Step 6: Let r offspring and s parent strings form the new population $P(t+1)$.

The selection in step 2,3 and 5 is implemented with respect to a probability distribution based on the strings' fitness. The probability for every string to be selected is calculated as follows:

(1) Calculate the fitness value $fit(s_i)$ for every string s_i, $i = 1, \ldots, popsize$. The fitness values are calculated by an information criterion (chapter 4). Fitness and information criterion are connected by mathematical transformations described in section 5.2.2.

(2) Determine the total fitness of the population

$$F = \sum_{i=1}^{popsize} fit(s_i) \, .$$

(3) Calculate the probability p_i and the cumulative probability q_i of a selection for each string s_i, $i = 1, \ldots, popsize$, by

$$p_i = \frac{fit(s_i)}{F} \, , \qquad q_i = \sum_{j=1}^{i} p_j \, .$$

To select a single string for the new population, the user first needs to generate *popsize* random (float) numbers $r_i \in [0, 1]$ and then check for every r_i, $i = 1, \ldots, popsize$:

- if $r_i \leq q_1$ then select the first string s_1;

- if $r_i \geq q_i$ then select the jth string s_i if $q_{j-1} < r_i \leq q_j$ where $j = 2, \ldots, popsize$.

Hence more fit strings have a larger probability to be chosen compared to the less fit strings.

Figure 5.8 presents the genetic algorithm, used in the simulation studies in chapters 6 and 7. Therefore our genetic algorithm has several build-in steps, which increase the effectiveness compared with many other conventional genetic algorithms:

- Deletion of a default number of worse strings in population $P(t)$ limits the group of strings available for future iterations and thus the selective pressure is high.

- Strings of a population $P(t)$ of high fitness will enter the new population $P(t+1)$ either as offsprings (step 2) or as parent (step 3) with high probability. By step 3 the best solutions of the old population are not forgotten.

- Exact copies of strings are not allowed. Hence there is no danger that a few strings (we call them *super-individuals*) generate many equal copies and thus repress other less fit strings. Mutation of some genes yields new strings of different genotype. The size of a string which will be mutated (and hence the size of lost original information) is controlled by a random number r.

  Prevention of several equal strings improves the diversity of a population. There will be only a slight increase of selective pressure if we change the genotype of a string by controlled mutation (because most strings maintain their original information).

- The classical mutation-step (Michalewicz (1996)) is canceled. Instead, only step 4 will prevent equal strings.

As termination condition we compare the fitness values of the fittest strings between two adjacent populations. The more fit value (and string) are saved. If the fitness does not change during a default number $term \in \{2, \ldots, T\}$ of successive iterations (T = maximal iteration number) the genetic algorithm is terminated. All simulations have $term = 20$.

For application of the genetic algorithm described above the following parameter must be chosen by the user: population size (*popsize*), crossover probability p_c, deletion of u percentage of worst strings, selection of r parents and s offsprings, iteration number T, parameter ν of the IAC operator and parameter b of the non-uniform mutation operator.

Remark 10. The software of the genetic algorithm for optimization of real-valued parameters presented above can be downloaded from internet. The address is *http://www.stat.uni-muenchen.de/sfb386/*. A program manual of this software, called *GENcon* is available in appendix A. For better illustration an example from the field of mathematical function optimization is given there.

5.4 Some theoretical Aspects of Genetic Algorithms

To evaluate the performance of an optimization algorithm the results received from practical problems are not sufficient. In fact we need some general results, which are independent of special problems. In particular the following questions are of interest:

- Why does the algorithm work?

- Does the algorithm converge?

- At what time does the user obtain adequate results (speed of convergence)?

Since the genetic algorithms are open to a public domain by the book of Goldberg (1989), many research groups are interested in generating a theoretical concept generally accepted. Even if the concept still shows diverse open questions, there has been clear progress (e.g. also in the questions formulated above). In this section we just give a short outline of the concepts, dealing with functionality, convergence and time complexity in genetic algorithms.

5.4.1 The Schema Theorem

For understanding the behavior and efficiency of genetic algorithms as a searching technique, Holland (1975) introduces the *schema*-concept yielding the basis for diverse theoretical considerations. Here the schema-concept is discussed for binary strings (compare Michalewicz (1996)) and the extension for differently coded strings will be sketched subsequently.

Definition 4. *A schema is a template describing a set of strings which agree in some defined positions. The remaining positions of the strings are not determined and become marked in the schema by a don't care symbol* $(\star)$.

Example 5. The schema $S_1 = (101\star\star)$ matches all binary strings of length 5, where the first three positions are fixed. The following set of strings is matched by schema S_1:

$$\{(10100),(10101),(10110),(10111)\}.$$

It is clear that the schema $S_2 = (10100)$ represents only one string respectively the schema $S_3 = (\star\star\star\star\star)$ represents all strings of length 5. Thus schema S_3 is the most general schema of length 5, because it contains no fixed genes.

Definition 5. *Let S a schema which contains a set of strings.*

(i) The order of a schema S is the number of fixed positions (i.e. number of 0 and 1). We denote the order as $o(S)$.

(ii) The defining length of a schema S is the distance between the first and the last fixed string position. We denote the defining length as $\delta(S)$.

Example 6. The schema $S_4 = (\star 101 \star \star 0)$ of length 7 has order $o(S_4) = 4$ and defining length $\delta(S_4) = 7 - 2 = 5$.

For derivation of the schema theorem and the building block hypothesis we start with a population at iteration step t. Let be $\xi(S,t)$ the number of strings in a population at time t matched by schema S. Another property is the fitness at time t, denoted as $fit_{avg}(S,t)$, and defined by

$$fit_{avg}(S,t) = \frac{1}{p}\sum_{i=1}^{p} fit(s_i,t)$$

where we have assumed that p strings $s_i, i = 1,\ldots,p$, in a population are matched by a schema S at time t.

During selection step an intermediate population is created. This population contains strings according to its fitness. As seen in section 5.3.4 a string has the probability $p_i = fit(s_i)/F(t)$ to be selected, where $F(t)$ is the total fitness of the whole population at time t. If we assume a selection of *popsize* strings, we expect

$$\xi(S,t+1) = \xi(S,t) \cdot popsize \cdot \frac{fit_{avg}}{F(t)} \qquad (5.9)$$

in the next iteration step $t+1$. Here the third factor of formula (5.9) yields the probability of the selection of such a schema string. We can rewrite

(5.9) by taking into account that the average fitness of the population is given as $\overline{F(t)} = F(t)/popsize$ and thus

$$\xi(S, t+1) = \xi(S, t) \cdot \frac{fit_{avg}}{\overline{F(t)}}. \tag{5.10}$$

The so called *reproductive schema growth equation* shows that the number of strings in the population grows as a ratio of the fitness of the schema to the average fitness of the population. Hence an "above-average" schema receives an increasing number of strings in the next generation, a "below average" schema receives a decreasing number of strings.

If we assume that schema S remains above-average by $\epsilon\%$ (i.e. $fit_{avg} = \overline{F(t)} + \epsilon \cdot \overline{F(t)}$) we get from formula (5.10)

$$\xi(S, t+1) = \xi(S, 0)(1 + \epsilon)^t$$

and hence we have not only an "above-average" schema, but, more precisely, a schema which receives an *exponentially* increasing number of strings in the next generation.

In the next step of the evolution cycle the crossover operator is applied. Generally the crossing point *pos* is uniformly selected among $m - 1$ positions of a string (compare also one-point crossover in example 3). Hence the probability of a schema survival is given as

$$p_s(S) = 1 - p_c \cdot \frac{\delta(S)}{m - 1}, \tag{5.11}$$

under the condition that only some strings, controlled by the crossover probability p_c participate in the crossover process.

Remark 11. Generally speaking, if a crossover point is selected between fixed positions in a schema it will be destroyed. Thereby the defining length plays a significant role in the probability of destruction and survival of strings: if the defining length $\delta(s)$ is small, we also have a small destruction probability and vice versa. But there is still a chance for the schema to survive, e.g. if both strings, participating in the crossover process, have the same form (this probability of such an event is quite small). Hence we should replace the equality in (5.11) by $\geq$.

Thus the reproductive schema growth equation can be extended by the expression (5.11)

$$\xi(S, t+1) \geq \xi(S, t) \cdot \frac{fit_{avg}}{F(t)} \cdot \left[1 - p_c \cdot \frac{\delta(S)}{m-1}\right],$$

which describes the expected number of strings matching a schema S in the next generation.

Finally the mutation operator is applied to the genetic algorithm. A schema survives in a mutation process only if all fixed positions remain unchanged. The mutation probability p_m yields the destruction probability of each bit and hence the survival probability of a schema with order $o(S)$ is given by

$$p_s(S) = (1 - p_m)^{o(S)}$$
$$\approx 1 - o(S) \cdot p_m \quad,$$

where the probability can be approximated if $p_m \ll 1$.

The combined effect of selection, crossover and mutation leads to the final form of the reproductive schema growth equation

$$\xi(S, t+1) \geq \xi(S, t) \cdot \frac{fit_{avg}}{F(t)} \cdot \left[1 - p_c \cdot \frac{\delta(S)}{m-1} - o(S) \cdot p_m\right]. \qquad (5.12)$$

This equation yields the expected number of strings matching a schema S in the next generation as a function of the actual number of strings matching a schema, the relative fitness of the schema, and its defining length and order. It is obvious that above-average schemata with short defining length and low order would be sampled at exponentially increased rates.

The final result of the reproductive schema growth equation (5.12) leads to the famous schema theorem by Holland (1975)

Theorem 2. *(Schema Theorem) Short, low-order, above-average schemata receive exponentially increasing trials in subsequent generations of a genetic algorithm.*

Hence the Theorem expresses that short, low-order, above-average schemata are more frequently chosen in subsequent generations. Furthermore crossover and mutation operators (which are responsible for greater variablility in the population) have low influence on short and low-order schemata. An immediate result of the schema theorem is that genetic algorithms explore the search space by short, low-order schemata (Holland (1975)):

Building Block Hypothesis. *A genetic algorithm seeks near-optimal performance through the juxtaposition of short, low-order, high-performance schemata, called building blocks.*

Although some research has been done to prove this hypothesis (Bethke (1980)), for most nontrivial applications we rely on empirical results.

In the course of time diverse approaches have been presented to extend the schema-concept to genetic agorithms with other representations (e.g. Antonisse (1989), Wright (1991), Eshelman & Schaffer (1993)). Of notable interest are the works of Radcliffe (1991a) and Radcliffe (1991b). Radcliffe's extension of the Holland's schema-concept enables the application to other representations than the classical binary representation. By introduction of arbitrary equivalence relations also general non-string representations are possible.

Radcliffe argues that general schemata have to be defined in such a way that they form equivalence classes among a given equivalence relation. Intuitively, the idea is that two strings η and ζ are equivalent if they have the same bits at the fixed positions (i.e. positions which have no "don't care" symbol $\star$). All strings which are equivalent together can be assigned to the same equivalence class.

Example 7. Two strings $\eta = (0101)$ and $\zeta = (1101)$ belong to the schema $(\star 1 \star 1)$. Further η and ζ are equivalent, because the bits at fixed positions (i.e. the first respectively third bit) are equal. $\xi = (\star a \star b)$ is the associated equivalence class of those strings which have $\eta_2 = \zeta_2 = a$ and $\eta_4 = \zeta_4 = b$.

For reasonable application of the schema theorem to general schemata (denoted as *formae*) the diverse members -also called *instances*- (of such a formae) should have similar fitness values. Now the main objective is the finding of operators which handle the formae effectively. Therefore Radcliffe suggests six design principles for building useful formae, string representations and gentic operators:

(1) *Minimal redundancy:* each member of the search space should be represented by only one string, i.e. the coding function ρ from the search space S into a space of strings C

$$\rho : S \longrightarrow C$$

should ideally be a bijection.

(2) *Correlation within formae*: some of the equivalence relations must relate strings with correlated performance. This means the instances of one formae should have similar fitness (at least for some formae).

(3) *Closure*: the intersection of any pair of compatible formae should be a formae itself. Two formae ξ and ξ' are denoted as *compatible*, if it is possible for a string to be an instance of both ξ and ξ', i.e.

$$\xi \ compatible \iff \xi \cap \xi' \neq \emptyset.$$

(4) *Respect*: the crossing of two instances of any formae should produce another instance of that formae.

(5) *Proper assortment*: given instances of two compatible formae, it should be possible to cross them to produce a child which is an instance of both formae.

(6) *Ergodicity*: for any given population it should be possible through a finite sequence of applications of the genetic operators, to access any point in the search space. (For example if the whole population has blue eyes, it must be possible to produce a brown-eyed child. The mutation operator usually ensures this.)

The approach of Radcliffe not only extends the schema theorem to general representations but also considers how operators of a genetic algorithm can adequately be designed. It should be noticed that independent of Radcliffe also Vose (1991) and Vose & Liepins (1991) have developed a similar approach.

5.4.2 Convergence of Genetic Algorithms

Two aspects of global optimization procedures are intensely researched: the convergence to a global optimum and the rate of convergence. In context with genetic algorithms several authors have explored the convergence with diverse restricting assumptions. The problem of convergence is influenced by different perspectives. Under restriction of only a finite population and no application of the crossover operator, Goldberg & Segrest (1987) provided a finite Markov chain analysis of genetic algorithm. Hartl (1990) and Fogel (1992) developed a possibility by using a Markov chain genetic algorithm model and a combination of the existing convergence theory on the simulated annealing algorithm. Eiben, Arts & Van

Hee (1991) introduced an abstract genetic algorithm which unifies genetic algorithms and simulated annealing. For that algorithm they discussed a Markov chain analysis and yield conditions for the convergence to a global optimum with probability 1. Rudolph (1994) showed that a classical genetic algorithm never converges to a global optimum. But that is not the case for genetic algorithms which use an elitist-strategy (i.e. the best solutions maintained in the population). A comparable result was published by Bhandari, Murthy & Pal (1996): they have proved that an elitist genetic algorithm with strings of fixed length converges to an optimal string if the number of iterations goes to infinity. Their only assumptions are: (1) the optimal string of the present population has a fitness value at least of the fitness of the optimal strings from the previous populations; (2) within any given iteration each string has positive probability to take the optimal string.

Another interesting approach to the proof of the genetic algorithms' convergence (without elitist model) is introduced by Michalewicz (1996) and is based on the Banach fixpoint theorem. The only restriction is that there should be an improvement of subsequent populations (not necessarily improvement of the best string). In the following we give a formal presentation of this concept.

Before application of the Banach fixpoint theorem on genetic algorithms we shortly specify some necessary definitions and basics around this theorem.

Definition 6. *Let R be a set of real numbers. A set M together with a mapping $\delta : M \times M \longrightarrow R$ is a metric space $\langle M, \delta \rangle$ if the following conditions are satisfied for any elements $x, y, z \in M$*

- $\delta(x, y) \geq 0$ *and* $\delta(x, y) = 0$ *if* $x = y$;

- $\delta(x, y) = \delta(y, x)$;

- $\delta(x, y) + \delta(y, z) \geq \delta(x, z)$.

The mapping δ is called distance.

Definition 7. *Let $\langle M, \delta \rangle$ be a metric space and let $f : M \longrightarrow M$ be a mapping. f is denoted as contractive if there is a constant $c \in [0, 1)$ such that for all $x, y \in M$ holds*

$$\delta(f(x), f(y)) \leq c \cdot \delta(x, y).$$

Definition 8. *Let $\langle M, \delta \rangle$ be a metric space.*

(i) The sequence $p_0, p_1, \ldots$ of elements is a Cauchy sequence if for any $\epsilon > 0$ there is a k such that for all $m, n > k$ holds

$$\delta(p_m, p_n) < \epsilon.$$

(ii) A metric space is complete if any Cauchy sequence $p_0, p_1, \ldots$ has a limit

$$p = \lim_{n \to \infty} p_n \ .$$

With these definitions it is possible to formulate the Banach fixpoint theorem. The proof can be found in most books on topology (e.g. Dixmier (1984)).

Theorem 3. *(Banach Fixpoint Theorem) Let $\langle M, \delta \rangle$ be a complete metric space and let $f : M \longrightarrow M$ be a contractive mapping. Then f has a unique fixpoint $x \in M$ such that for any $x_0 \in M$ holds*

$$x = \lim_{i \to \infty} f^i(x_0),$$

where $f^0(x_0) = x_0$ and $f^{i+1}(x_0) = f(f^i(x_0))$.

Now we have to apply the Banach fixpoint theorem to genetic algorithms. For that purpose it is possible to define genetic algorithms as transformations between populations. The idea is to construct a metric space which contains populations as elements. Then we have to find a contractive mapping f between genetic algorithms (i.e. between the populations of diverse genetic algorithms). Finally applying of the Banach fixpoint theorem yields the convergence of those algorithms.

The idea can be formulated more formally in the following theorem. As mentioned above we assume an improvement of the populations' fitness between succeeding iterations.

Theorem 4. *Let R be a set of real numbers. We assume that every population P consists of n strings $s_i, i = 1, \ldots, n$, i.e. $P = \{s_1, \ldots, s_n\}$. Moreover the fitness function for a population is given as*

$$fit_{avg}(P) = \frac{1}{n} \sum_{i=1}^{n} fit(s_i),$$

where $fit(s_i), i = 1 \ldots, n$ yields the fitness of the strings in population P. Let M be a set which consists of all possible populations P, i.e. any vector $\{s_1, \ldots, s_n\} \in M$. Then holds

(i) $\langle M, \delta \rangle$, consisting of a set M and a distance $\delta : M \times M \longrightarrow R$ is a metric space. The distance is defined as

$$\delta(P_1, P_2) = \begin{cases} 0 & if \ P_1 = P_2 \\ |1 + l_{up} - fit_{avg}(P_1)| + |1 + l_{up} - fit_{avg}(P_2)| & otherwise \end{cases}$$

where l_{up} is the upper limit of the fitness function, i.e. $fit(s_i) \leq l_{up}$ and thus $fit_{avg} \leq l_{up}$ for all individuals $s_i, i = 1, \ldots, n$. Furthermore the metric space $\langle M, \delta \rangle$ is complete.

(ii) A single iteration (consisting of selection, crossover, mutation and fitness calculation) of a run of genetic algorithm is a contractive mapping

$$f : M \longrightarrow M$$
$$P(t) \mapsto f(P(t)) = P(t+1),$$

if we have an improvement of the populations' fitness between succeeding iterations (without any loss of generality the t-th iteration), i.e.

$$fit_{avg}(P(t)) < fit_{avg}(P(t+1)). \tag{5.13}$$

Proof. (i) First we have to show the three properties of Definition 6.

- $\delta(P_1, P_2) \geq 0$ for any populations P_1 and P_2. Moreover if population $P_1 = P_2$, we have $\delta(P_1, P_2) = 0$;

- $\delta(P_1, P_2) = \delta(P_2, P_1)$;

- $\delta(P_1, P_2) + \delta(P_2, P_3) = |1 + l_{up} - fit_{avg}(P_1)| + |1 + l_{up} - fit_{avg}(P_2)| + |1 + l_{up} - fit_{avg}(P_2)| + |1 + l_{up} - fit_{avg}(P_3)| \geq |1 + l_{up} - fit_{avg}(P_1)| + |1 + l_{up} - fit_{avg}(P_3)| = \delta(P_1, P_3)$.

Hence $\langle M, \delta \rangle$ is a metric space.

Furthermore the metric space $\langle M, \delta \rangle$ is complete, because for any Cauchy sequence $P_1, P_2, \ldots$ of populations (a population has usually a finite number of elements) there exists a k such that $P_n = P_k$ for all $n > k$. Hence all Cauchy sequences P_i have a limit for $i \to \infty$.

Remark 12. In case of genetic algorithms we generally work with finite metric spaces, because a population only contains a finite number of strings. Thus the demand of Banach for completeness of metric spaces in that case is always satisfied.

(ii) A single iteration is a contractive mapping $f : M \longrightarrow M$ because with (5.13) holds

$$
\begin{aligned}
\delta(f(P_1(t)), f(P_2(t))) = & \\
= & \ |1 + l_{up} - \underbrace{fit_{avg}(f(P_1(t)))}_{> fit_{avg}(P_1(t))}| + |1 + l_{up} - \underbrace{fit_{avg}(f(P_2(t)))}_{> fit_{avg}(P_2(t))}| \\
< & \ |1 + l_{up} - fit_{avg}(P_1(t))| + |1 + l_{up} - fit_{avg}(P_2(t))| \\
= & \ \delta(P_1(t), P_2(t)).
\end{aligned}
$$

Remark 13. If there is no improvement, we do not count such iteration and thus we repeat selection, crossover, mutation and fitness calculation.

$\square$

With the results presented so far we can apply the Banach fixpoint theorem in context with genetic algorithms and obtain the convergence of those algorithms:

Theorem 5. *Let $\langle M, \delta \rangle$ (space of populations) be a complete metric space. The iteration $f : P(t) \longrightarrow P(t+1)$ is a contractive mapping and improves the populations' fitness. Then f has a unique fixpoint $P^\star \in M$ such that for any $P(0) \in M$*

$$
P^\star = \lim_{i \to \infty} f^i(P(0)),
$$

where P(0) is the initial population.

Hence the genetic algorithm converges to population $P^\star$ which is a unique fixpoint in the search space of all populations. Because of

$$
fit_{avg}(P) = \frac{1}{n} \sum_{i=1}^{n} fit(s_i),
$$

the fixpoint $P^\star$ is achieved only if all strings in the population have the same global value. It is obvious that $P^\star$ does not depend on the initial population $P(0)$.

In case of a contractive mapping genetic algorithm we always have convergence to a global optimum (in infinite time). Here the initial population influences the convergence speed, only. The main problem resulting from this convergence is the possibility that the genetic algorithm yields no fitness improvement during a long number of iterations. That means operators like crossover and mutation are unable to produce a better (fitter) population and thus the algorithm is getting stuck in a suboptimal solution for a long time. Generally the user is not interested in proofs of convergence. He is more interested in the performance of solution, temporarily achievable, because computer capacity and running time of a genetic algorithm are not available unlimited. Hence time complexity is an important measure for evaluation quality optimization of an algorithm.

An very interesting result concerning time complexity can be found in Hart & Belew (1991). They analyse the class $\mathcal{F}$ of all deterministic pseudo-boolean functions, i.e. functions of the form

$$f : \{0, 1\}^l \longrightarrow \mathbb{R},$$

where l is the number of bits in a string. For each function in $\mathcal{F}$ it is not possible to find any genetic algorithm, yielding a result in polynomial time that differs less than a constant $\delta \in \mathbb{R}^+$ from the global optimum. In other words there are functions in $\mathcal{F}$ for which the genetic algorithm is a poor optimizer. From these results we can follow that besides population size and parameters of the operators (e.g. crossover probability), also the class of fitness functions used have to be specified.

Deb & Goldberg (1991) analyse the influence of the selection procedure on the speed of convergence of genetic algorithms. With the assumption that only the selection procedure changes the population, they obtain the following complexity times:

- stochastic universal sampling needs about $O(n \, log \, n)$ iterations;

- rank-based respectively tournament selection needs about $O(log \, n)$ iterations.

Because of the restricted assumption these results can only be applied as a tendency to general genetic algorithms.

For estimation of time complexity for arbitrary functions, Louis & Rawlins (1992) use the average Hamming distance in the population during the run of genetic algorithms.

Definition 9. (Hamming distance) *Let s_1 and s_2 be two strings of length l. Then the Hamming distance $d(s_1, s_2)$ is defined as*

$$d(s_1, s_2) = \sum_{i=1}^{l} |s_{1_i} - s_{2_i}| \in [0, l],$$

i.e. the number of differences between respective bits of the strings s_1 and s_2 are counted.

Assuming that similar strings also have similar fitness values and the mutation operator has no influence, the average Hamming distance d_t at iteration t is given by

$$d_t = \begin{cases} \frac{l}{2} & \text{if } t = 0 \\ a^t \cdot h_0 & \text{if } t > 0 \end{cases}, \tag{5.14}$$

where a is estimated by the Hamming distance change during the first ten iterations. By comparison of theoretical and simulated Hamming distances for diverse test functions they found a close agreement between theoretical and predicted results.

A drawback is that the standard deviation of the simulated Hamming distances averages more than 50% and hence the prognosis of formula (5.14) is restricted. Furthermore the estimation of a is very extensive in case of large populations.

6

Simulation Study

In this chapter the performance of the genetic algorithm for smoothing parameter selection is analysed and compared with other methods related in literature. The methods we use in the following simulations are sketched in section 6.1. As program packages for these methods do not often use the additive structure, we first compare them in section 6.2 for the single covariate case with functions of rather different spatial variability. Then section 6.3 compares our approach with other methods in literature by means of a simulated additive model.

Remark 14. It should be noted that the software tool for smoothing parameter optimization, called *SMAD* (*SM*oothing in *A*dditive *M*odels), is based on the genetic algorithm *GENcon* (appendix A). The manual of SMAD is given in appendix B and the software is available on internet (*http://www.stat.uni-muenchen.de/sfb386/*).

6.1 Alternative Approaches

This section briefly describes alternative approaches to estimate functions and to select smoothing parameters which are compared with the present approach. All approaches are based on the expansion in basis functions with the predictor

$$\eta(\mathbf{x}_i) = \beta_0 + \sum_{j=1}^{p} \sum_{\nu=1}^{K_j} \beta_{j\nu} \phi_{j\nu}(x_{ij}). \tag{6.1}$$

6.1.1 Mixed Models

An approach based on the methodology of mixed models has been used by Parise, Wand, Ruppert & Ryan (2001). The basic concept is to treat the parameters in (6.1) as random effects. With respect to that strategy and in context with additive models with truncated power series $\phi_{j\nu}(x) = (x - \xi_{j\nu})_+$ as basis functions it can be assumed that

$$\beta_{j\nu} \sim \mathcal{N}(0, \sigma_j^2), \quad \nu = 1, \dots, K_j,$$

with $\beta_{j1}, \dots, \beta_{jK_j}$ being independent. The σ_j^2 express variability of the parameters: $\sigma_j^2 = \infty$ corresponds to the unrestricted case whereas $\sigma_j^2 \to 0$ implies high restrictions of the parameters. The estimation of structural and smoothing parameters is based on solving the generalized mixed models' equation. In the simulation study we assume that adjacent weights of the truncated power series basis used are correlated with a first order autoregressive (AR(1)) structure whose parameters are automatically estimated by the software package SAS used.

6.1.2 Bayesian P-splines

A fully Bayesian approach has been used by Lang & Brezger (2004). In a similar way as with the mixed model approach, the parameters are considered as random. In context with basis functions as B-splines one assumes prior distribution on the parameters. This may be considered as the stochastic analogue to the use of a penalty term in the estimation procedure. For the first order differences one assumes diffuse priors for β_{j1} and a first order random walk $\beta_{j\nu} = \beta_{j,\nu-1} + u_{j\nu}$ with Gaussian errors $u_{j\nu} \sim \mathcal{N}(0, \sigma_j^2)$. For full Bayesian inference, hyper priors are assigned to the parameters σ_j^2, using highly dispersed inverse Gamma priors, $p(\sigma_j^2) \sim \mathcal{IG}(a_j, b_j)$ with a_j, b_j fixed. Lang & Brezger (2004) use $a_j = 1, b_j = 0.005$. Inference is based on Markov Chain Monte Carlo (MCMC) simulation techniques.

If one uses B-splines which may be constructed from the truncated power series the assumption of independent random effects is replaced by assuming that differences of parameters are normally distributed. In the simpler case one assumes $\beta_{j,\nu+1} - \beta_{j\nu} \sim \mathcal{N}(0, \sigma_j^2)$.

6.1.3 Relevance Vector Machine

The relevance vector machine (Tipping (2000), Tipping (2001)) has been developed by the machine learning community as an improvement of the support vector machine. Tipping also uses a Bayesian framework. Starting with one basis function at each observation the weights $\beta_{j\nu}$ are independent and normally distributed, $\beta_{j\nu} \sim \mathcal{N}(0, \alpha_i^{-1})$ whereas the hyperpriors for α_i and σ^2 are gamma distributions optimized by a marginal likelihood approach. The essential difference between Tipping's algorithm and Bayesian approaches is that the number of basis functions initially equals to the number of observations. Then it reduces to only few remaining basis functions. For the rest the weights become zero. In our simulation study (section 6.2) we use 40 respectively 80 Gaussian kernels as basis functions with different σ_{gauss}, in detail $\sigma_{gauss} = 0.15/\sqrt{2}$ (function with $j = 3$) and $\sigma_{gauss} = 0.06/\sqrt{2}$ (function with $j = 6$). The hyperpriors are automatically estimated by the software program SAS.

6.1.4 Adaptive Regression

Friedman (1991) proposed *multivariate adaptive regression splines* in short *MARS*. MARS uses the expansion of basis functions of the form

$$\eta(\mathbf{x}_i) = \beta_0 + \sum_{j=1}^{p} \sum_{\nu=1}^{K_j} \beta_{j\nu} \phi_{j\nu}(x_{ij})$$

in a stepwise way where basis functions are successively constructed from products of linear splines $(x_{ij} - \xi_{j\nu})_+$ and $(\xi_{j\nu} - x_i)_+$, where x_{ij} is one component of the vector $\mathbf{x}_i$ and $\xi_{j\nu}$ are knots chosen from the observation of the corresponding component of $\mathbf{x}_i$. By stepwise inclusion of linear splines a large model (we use a maximum number of 150 basis functions) is obtained for which a backward deletion procedure is often applied. For details see Friedman (1991).

An alternative Bayesian procedure applying adaptive regression splines has been proposed by Biller & Fahrmeir (2002). Like other Bayesian approaches this procedure is not a stepwise approach and is based on a large set of basis functions like B-splines, characterized for each variable by candidate knots. In addition to the parameters, the number of knots as well as the specific choice of knots are specified by prior distributions. For estimations, the number of knots are Poisson distributed with a mean number of 20 knots.

6.1.5 Smoothing Parameter Selection with S-Plus-Software

The software package S-Plus offers a restricted possibility of smoothing parameter selection. First one calculates AIC for an initial model. Then one has to specify a list with other modelling alternatives. Each covariate can be dropped or integrated in a model as a linear term respectively as a B-spline with a default penalty term. Therefore Eilers & Marx (1996) published a S-Plus-function allowing the expansion of each function $f_j, j = 1, 2, \ldots$ in B-splines with penalty term. Starting with the initial model the implemented function `step` successively calculates the AIC for all alternative models. If a current model yields a better AIC we replace the previous model. Because of its implementation S-Plus can only run a relatively small number of different models. In the simulation study of section 6.3 it has been shown, that for an additive model of 5 functions $f_j, j = 1, \ldots, 5$, where each one is expanded in 20 B-splines, we can select from a list of about 17 models (i.e. each covariate can be modelled linearly or as a B-spline with one of 16 different smoothing parameters). However, to present results of the popular statistical software tool, we had to make assumptions concerning the choice of smoothing parameters. Hence a grid of 16 smoothing parameters was log-spaced between 10^{-2} and 10^2, i.e. their base-10 logarithms were equally spaced between -2 and 2. It is obvious that this discrete and restricted smoothing parameter selection yields inexcact solutions and that the optimal choice of smoothing parameters is very rare. Furthermore, we do not usually know the function's true structure and hence the restrictions mentioned above are not generally supportable.

6.1.6 Smoothing Parameter Selection with R-Software

The statistic software package `mgcv` (Wood (2001)) running in R contains an automatic smoothing parameter selection, based on a method first proposed by Gu & Wahba (1991). The idea is to re-write the multiple smoothing parameter model fitting problem with an extra "overall" smoothing parameter controlling the tradeoff between model adaption and overall smoothness. The smoothing parameters retained now control only the relative weights given to the different penalty terms. Then, the approach is to alternate the following steps:

- Estimation of the overall smoothing parameters using one-dimensional direct search methods.

- Update the relative smoothing parameters simultaneously by using the Newton method.

The approach is based on minimizing the Generalized Cross Validation (GCV) as model selection criterion. In our simulations we use cubic B-splines whereby the number of knots can be adjusted by hand. For further details see Wood (2000) and Wood (2001).

6.2 Estimation of different Oscillating Functions

In our first simulation we consider the function

$$f(x) = \sqrt{x(1-x)} \sin \left(\frac{2\pi(1 + 2^{(9-4j)/5})}{x + 2^{(9-4j)/5}} \right).$$

The spatial variability of $f(x)$ can be changed by the parameter $j = 1, 2, \ldots$ (see Ruppert & Carroll (2000)). We simulate 250 datasets for low spatial variability ($j = 3$) or for high spatial variability ($j = 6$), respectively. Each dataset consists of 400 independently and uniformly distributed data with $\sigma = 0.2$ (see Figure 6.1). For estimating the function, $f(x)$ is expanded in 40 (respectively 80) cubic B-spline basis functions. As penalty we use the third order differences of adjacent coefficients and the smoothing parameter chosen from the interval $[10^{-4}, 10^4]$. The default parameters of the genetic algorithm used are: population size (*popsize*) = 48 strings, crossover probability $p_c = 0.5$, deletion of $u = 60$ percentage of worst strings, selection of $r = 30$ and $s = 18$ strings, $\nu = 0.5, T = 1000$ and $b = 1$.

To compare our approach with other methods we computed $log_{10}(\sqrt{MSE})$ with empirical mean squared error given by

$$MSE(\hat{f}) = \frac{1}{n} \sum_{i=1}^{n} (f(x_i) - \hat{f}(x_i))^2.$$

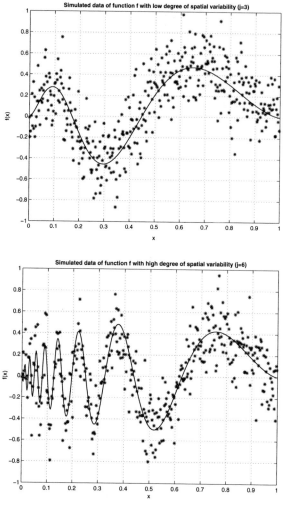

Figure 6.1. *The panels show the true functions (solid line) with spatial variability $j = 3$ (top) and $j = 6$ (below) for a randomly chosen dataset with $\sigma = 0.2$.*

For both specifications of spatial variability ($j = 3, 6$) Figure 6.2 shows boxplots of $log_{10}(\sqrt{MSE})$ for various estimators. Here only one global smoothing parameter is used. From left to right the boxplots refer to genetic algorithm (40 and 80 knots), Relevance Vector Machine (RVM, 40 and 80 knots), mixed model (40 and 80 knots), Bayesian adaptive regres-

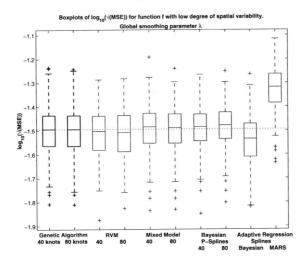

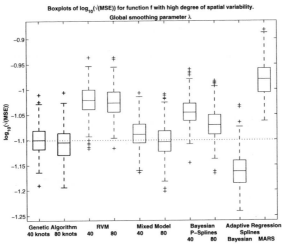

Figure 6.2. *The panels show boxplots of $\log_{10}(\sqrt{MSE})$ for various estimators for $j = 3$ (top) and $j = 6$ (below). The dotted line represents the median of the genetic algorithm with 40 knots.*

sion splines, and MARS. For better comparison the dotted line represents the median of the genetic algorithm with 40 knots.

From Figure 6.2 we can draw the following conclusions:

- For $j = 3$ most approaches yield similar results. The MARS approach leads to substantially poorer results than all the other methods. For $j = 6$ the performance strongly depends on the method. While mixed models approximately yield the same results as the genetic algorithm, RVM, Bayesian P-splines and MARS show poorer results. Only Bayesian adaptive regression splines lead to better results compared to the genetic algorithm.

- For $j = 3$ doubling the number of basis functions from 40 to 80 knots scarcely improves the performance of the estimators. In case of Bayesian P-splines they even show a slight deterioration. For $j = 6$ an increasing number of basis functions yield better estimators on an average. But the magnitude of the improvements depends on the type of the approach.

For both specifications ($j = 3, 6$) Figure 6.3 yields the boxplots log_{10} ($\sqrt{MSE}$) of the simulation used above (i.e. equal datasets and default parameters of the genetic algorithm). But now we use local smoothing parameters in the interval $[10^{-4}, 10^4]$.

For direct comparison we present the results of our approach and Bayesian P-splines:

- Both specifications ($j = 3, 6$) yield comparably good estimators.

- The use of genetic algorithms with local smoothing parameters shows no improvements compared to genetic algorithms with global smoothing parameters.

To gain better insight into the estimation by means of genetic algorithms with global and local smoothing parameters, we have a look at Figure 6.4. For both specifications of spatial variability, simulation was run with 40 B-splines for quadratic (degree = 2) and cubic (degree = 3) B-splines and different penalty (order = 1, 2, 3). The dotted line represents the median of cubic B-splines with penalty order 3 which we used with the simulations above.

- For $j = 3$ an increasing penalty order improves the performance of the estimator. Apart from penalty order 1, the degree of B-splines has little influence.

- For $j = 6$ the cubic boxplots show more accurate estimators compared with the quadratic boxplots. In contrast to $j = 3$ the estimators become worse with increasing penalty order.

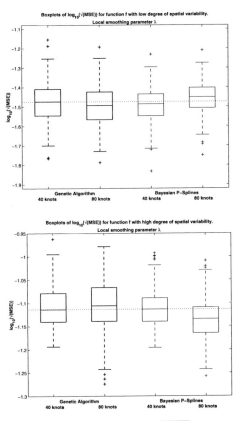

Figure 6.3. *The panels show boxplots of $\log_{10}(\sqrt{MSE})$ of the genetic algorithm and Bayesian P-splines with local smoothing parameters for $j = 3$ (top) and $j = 6$ (below). The dotted line represents the median of the genetic algorithm with 40 knots.*

These results confirm, that B-spline degree and penalty order affect the performance of an estimator. For different oscillating functions, however, choosing a suitable degree of B-splines and penalty order becomes more difficult. For example, our choice of cubic B-splines with penalty order 3 is suitable for low spatial functions. But a lower penalty order seems to be more accurate for highly oscillating functions. Comparable simulations with local smoothing parameters yield similar results as described in Figure 6.4.

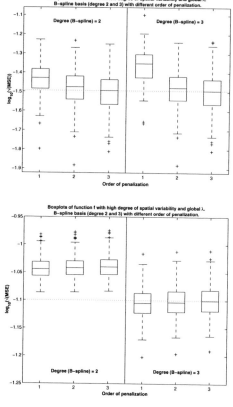

Figure 6.4. *Results for the genetic algorithm with global smoothing parameters but different degrees of B-spline and penalty order for the functions $j = 3$ (top) and $j = 6$ (below).*

In general, the underlying functions (e.g. in an additive model) are completely unknown and thus we have no idea about the degree of spatial variability. The choice of cubic B-splines with penalty order 2 or 3 should be an adequate solution for all kinds of functions.

6.3 Estimation for Additive Models

In this simulation we choose an additive model, consisting of 5 functions $f_j(x_{ij}), j = 1, \ldots, 5$ (Figure 6.5). We simulate 250 datasets, where each

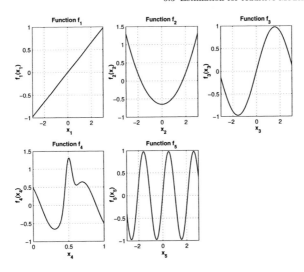

Figure 6.5. *Here the five original functions of the additive model used are shown.*

dataset consists of 500 independently and uniformly distributed data with $\sigma_1 = 0.3$ and $\sigma_2 = 0.6$. To estimate the single functions $f_j(x_{ij})$ we expand each function into 20 cubic B-spline basis functions. For penalty we use the third difference of adjacent coefficients. The five global smoothing parameters can be chosen from the interval $[10^{-4}, 10^4]$. The default parameters of the genetic algorithm are the same as in section 6.2.

To compare the results of our approach with other methods we computed $log(MSE)$. Figures 6.6 and 6.7 shows:

- In both cases the linear function f_1 is estimated best by S-Plus.

- Function f_2 is significantly better estimated for $\sigma_1 = 0.3$ and $\sigma_2 = 0.6$ by the genetic algorithm than all the other approaches.

- Apart from S-Plus, in case of $\sigma_1 = 0.3$ function f_3 is similarly estimated by all approaches. But in the case $\sigma_2 = 0.6$, the genetic algorithm yields better results compared with the other approaches.

- Bayesian adaptive regression splines significantly yields the best results in estimation of function f_4. For $\sigma_1 = 0.3$ the fit of function f_4 by the genetic algorithm is worse than that of all other approaches. But for

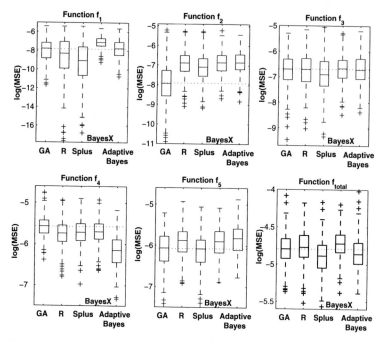

Figure 6.6. *Here boxplots of $\log(MSE)$ for the function $f_j, j = 1, \ldots, 5$ and the total function f_{total} of several estimation approaches with $\sigma_1 = 0.3$ are shown.*

$\sigma_2 = 0.6$ the approaches (except Bayesian adaptive regression splines) yield comparable results.

- Together with S-Plus, the genetic algorithm shows the best estimators of function f_5 for both specifications ($\sigma_1 = 0.3$ and $\sigma_2 = 0.6$).

- For $\sigma_1 = 0.3$ only S-Plus and Bayesian adaptive regression splines outperform the genetic algorithm in estimation of the total function f_{total}. But if we choose $\sigma_2 = 0.6$, the genetic algorithm estimates f_{total} better compared with all other approaches.

The simulation study shows that the results of the estimation of function f_{total} by Bayesian adaptive regression splines are closely connected to the fit of function f_4. Although this approach yields average results for all other functions, the excellent estimation of f_4 strongly influences the performance of the total function f_{total}. The reason is that variable knot selection of Bayesian adaptive regression splines adapts to the dif-

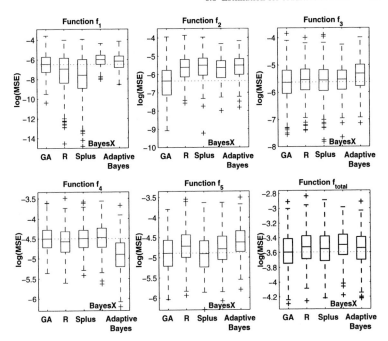

Figure 6.7. *Here boxplots of $log(MSE)$ for the function $f_j, j = 1, \ldots, 5$ and the total function f_{total} of several estimation approaches with $\sigma_2 = 0.6$ are shown.*

ferent spatial variability of function f_4. We notice again that only strict constraints of the smoothing parameter choice (section 6.1.5) lead to the results of the S-Plus approach. Furthermore, the results of function f_1 show the advantage of S-Plus to estimate the function f_1 by linear terms.

Table 6.1 shows the average running times per dataset of the used programs in the simulation of the additive model with $\sigma_1 = 0.3$. S-Plus, BayesX

Program	Running time in seconds
Genetic algorithm	337
R-package mgcv	3
S-Plus	468
BayesX	49
Adaptive Bayes	394

Table 6.1. *Here the average running times for estimation of one dataset in the simulation of an additive model with $\sigma_1 = 0.3$ are shown.*

and the R-package `mgcv` are commercial software and hence optimized with respect to running time. Genetic algorithm and adaptive Bayesian algorithm are primarily research tools, programmed for comparison of results and thus not optimized for running time. Hence a direct comparison is difficult. But Table 6.1 shows one obvious fact: the commercial software S-Plus uses much more running time to yield accurate results than all the other programs.

7

Effect of Fine Tuning on Smoothing Parameter Choice

Execution of appropriate optimization of smoothing parameters involves an accurate choice of parameters characterizing the genetic algorithm. Like other optimization algorithms, the genetic algorithm also contains parameters to be chosen before application. On the one side an appropriate parameter choice can lead to a better prediction accuracy to the true (but usually unknown) function. Otherwise simulation time (i.e. number of iterations up to convergence of the algorithm) can be decreased.

This chapter is organized as follows: in the next section we check the prediction accuracy and the simulation time for different values of diverse parameters, characterizing the genetic algorithm. In section 7.2 we deal with the question which influence the choice of an information criterion of chapter 4 have on simulated datasets. Finally section 7.3 briefly refers to the improved arithmetical crossover (IAC) operator presented in section 5.3.2 comparing this with the arithmetical crossover operator (Michalewicz (1996)).

Remark 15. The simulations in this chapter are the same as in section 6.2. Again we use 250 datasets for the function with low spatial variablity ($j = 3$) respectively for high spatial variability ($j = 6$). The functions are plotted in Figure 6.1. For estimating the function, $f(x)$ is expanded again in 40 cubic B-spline basis functions. But as penality we use second order differences of adjacent coefficients. Except the analysed parameter, the rest of parameters is chosen equally to section 6.2.

7.1 Genetic Algorithm with varying characteristic Parameters

Beside other characteristics the genetic algorithm presented in section 5.3 has the following parameters:

- parameter r describes the number of strings in a population used for the crossover process (r-portion). Hence s strings are not used for crossover;

- crossover probability p_c;

- size of population (*popsize*);

- proportion u of strings in a population which are deleted;

- influence of the constant for fitness calculation to be chosen in the transformation formula (compare 5.2.2).

As mentioned above, an appropriate choice of the characteristic parameters can result in a better prediction accuracy to the true function and a lower simulation time. As a measure for prediction accuracy we can use the empirical mean squared error (MSE)

$$MSE(\hat{f}) = \frac{1}{n} \sum_{i=1}^{n} (f(x_i) - \hat{f}(x_i))^2,$$

where we always calculate $log(MSE)$ in the following simulations.

In a first simulation we check the influence of diverse r/s-combinations on prediction accuracy and simulation time. The whole population $r + s = 48$ strings is divided into several different portions of r respectively s (compare Table 7.1). The results of the functions with low and high spatial variablity are plotted in Figure 7.1. Thereby the plots representing prediction accuracy yield similar results for all choices of the parameters r and s, i.e. all boxplots have comparable structure. Concerning simulation time we observe differences between the diverse parameter choices: a larger value of r (that is equipollent with the fact that a larger number of strings

Number of strings for r	10	18	28	38
Number of strings for s	38	30	20	10

Table 7.1. *Used r/s-combinations to check the influence of different portions of r respectively s on prediction accuracy and simulation time.*

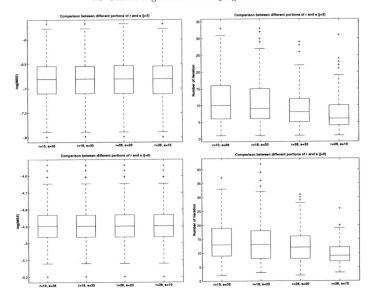

Figure 7.1. *The plots show the dependence of prediction accuracy (left plots) and simulation time (right plots) on the choice of the parameters r and s. The plots in the first row yield the results of the function with low spatial variability $(j = 3)$. The plots below show the respective results of the function with high spatial variability $(j = 6)$.*

is involved in the crossover process) leads to a decreasing number of iterations and thus there is a faster convergence of the genetic algorithm.

To check the influence of the crossover probability p_c, we have chosen again the parameters r and s as in section 6.2 (i.e. $r = 30, s = 18$). Variation of the value $p_c \in \{0.1, 0.3, 0.5, 0.7, 0.9\}$ yields the plots in the first row of Figure 7.2 showing prediction accuracy and simulation time in dependence of crossover probability p_c on the function of high spatial variability $(j = 6)$. Concerning prediction accuracy and simulation time, the boxplots show no significant variations. The reason for the negligible differences of simulation times might be the randomness of the genetic algorithm (e.g. generation of initial population is random). Analogous results as in Figure 7.2 can also be simulated for the function of low spatial variability $(j = 3)$.

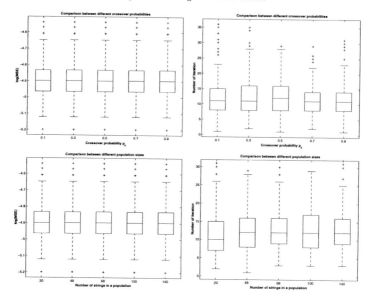

Figure 7.2. *Here the dependence of prediction accuracy and simulation time on different crossover probability values p_c is shown by the plots above. The plots below yield the comparable dependence on population size. All results have been simulated for the function with high spatial variability ($j = 6$).*

Remark 16. If the simulations of the functions of low and high spatial variablity (i.e. $j = 3, 6$) yield comparable results, we restrict ourselves (without loss of generality) to the case $j = 6$.

The plots of Figure 7.2 below show how prediction accuracy and simulation time depend on population size. For simulations we have chosen different numbers of strings in a population, in fact *popsize* $\in \{20, 48, 68, 100, 140\}$. Again the prediction accuracy of function $j = 6$ does not depend on population size. But for the lowest number of strings, simulation time leads to faster convergence of the algorithm. From a certain population size it is observable that there is no slower convergence for function $j = 6$ (respectively $j = 3$), i.e. the simulation time for diverse population sizes is similar.

The next simulation checks the dependence of prediction accuracy and simulation time on the proportion u of strings which have to be deleted. Therefore we have successively deleted $u \in \{10, 30, 60, 90\}$ percent of the

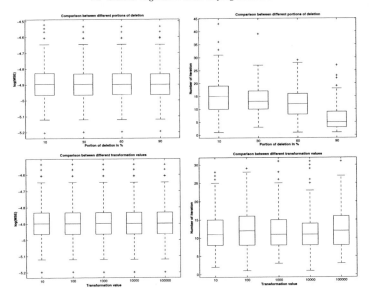

Figure 7.3. *The plots above show the dependence of prediction accuracy and simulation time on the deletion parameter u. The other plots yield the results for diverse transformation constants. All simulations are based on the function with high spatial variablity.*

worst strings in a population. The results are shown in Figure 7.3. The prediction accuracy has negligible differences for various values of parameter u. But there is an obvious decrease of simulation time if the proportion u of deleted strings increases. For example the genetic algorithm with $u = 90$ percent approximately takes one third of the simulation time, only, compared with $u = 10$ percent. Again these results can be calculated for function $j = 3$.

In section 5.2.2 we noted that transformation of fitness depends on a constant $\in \mathbb{R}$. The panels below on Figure 7.3 show that prediction accuracy and simulation time are independent of the size of the transformation constant chosen $\in \{10, 100, 1000, 10000, 100000\}$. The reason for the negligible differences of simulation times might be randomness of the genetic algorithm again. Similar results can be taken by application of function $j = 3$.

Thus the following results can be summarized from the simulations described above:

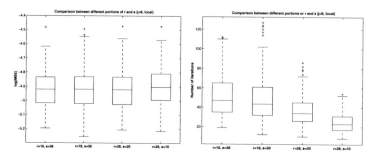

Figure 7.4. *For the function with high spatial variability the plots show the dependence of prediction accuracy respectively simulation time on different choices of parameters r and s.*

- an increasing portion r of strings (used for crossover process), reduces the simulation time;

- an increasing crossover probability p_c yields no changes of prediction accuracy and simulation time;

- a larger proportion u of deleted strings (and hence an increase in selective pressure) reduces simulation time;

- a low population size tends to faster convergence of the genetic algorithm. But from a certain population size it is observable that there is no slower convergence.

The simulations considered so far based on functions with one covariate and one (global) smoothing parameter which had to be optimized. If we simulate functions with several covariates respectively several (local) smoothing parameters, the differences between prediction accuracy and simulation time compared with different characteristic parameters are increasing. Especially longer simulation times can lead to higher computation costs. An example of an increase in simulation time is represented in the left plot of Figure 7.4. For the case of the function with high spatial variablity and local smoothing parameters (40 smoothing parameters, which have to be optimized), the influence of various r/s-combinations on simulation time are checked. In case of only one covariate and global smoothing parameter the various r/s-combinations differ at most in 4 iterations (compare Figure 7.1). But the highest difference in the current simulation shows 25 iterations. Hence an improper choice of characteristic parameters can result in a convergence entering into a significantly higher number of iterations. In the worst case of improper characteristic parameters it is possible that a global optimum may not be achieved, because the genetic algorithm terminates.

Remark 17. It should be pointed out that in the simulation with several covariates respectively smoothing parameters the prediction accuracy also shows some differences between the various r/s-combinations (compare Figure 7.4).

7.2 Choice of an appropriate Information Criterion

In addition to the choice of characteristical parameters in the genetic algorithm, an appropriate application of information criteria can also influence estimation quality respectively simulation time. In chapter 4 diverse information criteria have been presented and compared in a more theoretical way. In this section estimation quality (described by $log(MSE)$) and simulation time of these information criteria are compared within simulations.

For the function of different spatial variablity, Figure 7.5 shows the prediction accuracy of the true function and simulation time up to the termination of the genetic algorithm. The parameters of the genetic algorithm have been taken out of the simulation study of section 6.2.

Comparison of prediction accuracy for diverse information criteria shows little differences for both functions $j = 3, 6$. Only the boxplots of the BIC criterion tend to higher MSE values compared with the other criterions, i.e. there is lower prediction accuracy to the true function. Also the simulation times for the functions $j = 3, 6$ yield no clear winner. Maybe in case of the function with low spatial variability it seems that the improved AIC criterion tends to lower simulation times.

Similar prediction accuracy between different information criteria is shown in Figure 7.6 again. Here the average estimators of 250 datasets are plotted for all information criteria. We realize that the average curve of the BIC criterion yields the worst prediction compared with the other criteria. On the other side all information criteria except the BIC criterion have comparable structure.

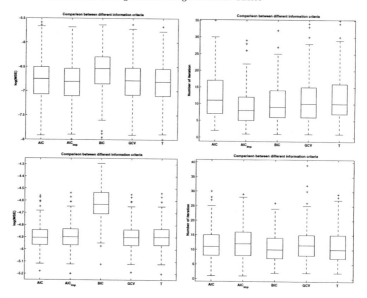

Figure 7.5.   *Here for the function with different spatial variability the dependence of prediction accuracy and simulation time on diverse information criteria is shown. The plots above yield the results for $j = 3$, the other ones for $j = 6$.*

Remark 18. Simulations for both functions $j = 3, 6$ and local smoothing parameters yield comparable results. Again the BIC criterion has the lowest prediction accuracy.

From the simulations presented above we can summarize that the diverse information criteria yield marginal differences in prediction accuracy and simulation time. Only the BIC criterion is an exception: in all simulations this criterion results in a significantly lower prediction accuracy compared with the other information criteria.

In context with variable selection we check again the application of diverse information criteria in simulations.

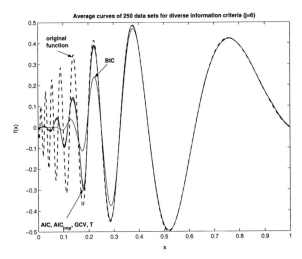

Figure 7.6. *For all information criteria the average of the estimators for* 250 *datasets are shown.*

7.3 Behavior of the Improved Arithmetical Crossover Operator in Simulations

We briefly refer again to the *improved arithmetical crossover (IAC)* operator presented in section 5.3.2. For the function of high spatial variability we compare this crossover operator with the arithmetical crossover operator (section 5.3.2).

The left plot of Figure 7.7 shows boxplots of $log_{10}(\sqrt{MSE})$ for the function of high spatial variability and global smoothing parameter. IAC yields estimators which are significantly better than the arithmetical crossover.

Moreover, IAC has faster speed of convergence (see Figure 7.7 (right)). Here, the mean $log_{10}(\sqrt{MSE})$ of all data sets for both crossover operators is shown. They were chosen by the genetic algorithm up to iteration $t = 200$. In each data set, the current population $P(s = t)$ of iteration t yields the $log_{10}(\sqrt{MSE})$-value only in the case where all former populations $P(s < t)$ have worse fitness. Otherwise, the $log_{10}(\sqrt{MSE})$-value of the former population $P(s = t - 1)$ is retained. For the curves in Figure 7.7 we average across the $log_{10}(\sqrt{MSE})$-values of the 100 data sets and realize a convergence to a minimal value after a few iterations if using IAC.

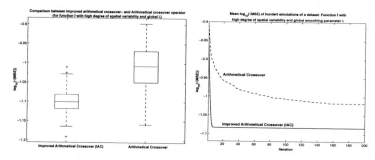

Figure 7.7.   *The left plot shows boxplots of $\log_{10}(\sqrt{MSE})$ of the genetic algorithm with improved arithmetical crossover and arithmetical crossover. We used 100 datasets of the high oscillating function $(j = 6)$ and a global smoothing parameter. The plot on the right side presents the mean $\log_{10}(\sqrt{MSE})$ of the data sets for both crossover operators.*

However, in general the arithmetical crossover operator does not obtain this minimal value even if we have a larger iteration number.

Part III

Variable Selection with Genetic Algorithms

Variable Selection in Additive Models

In many statistical applications (e.g. analysis of gene expression data) the user is often confronted with datasets having large numbers of explanatory variables. Although computers with increasing computing power have been developed in recent years, many datasets cannot be processed with common computers. Hence adequate variable selection procedures are necessary.

The problem of variable selection (or subset selection) arises when the relationship between a response variable and a subset of potential explanatory variables is to be modelled, but there is a substantial uncertainty about the relevance of the variables. Most datasets contain explanatory variables which are redundant or irrelevant and the objective is to eliminate these variables from datasets.

In the literature several algorithms for variable selection are proposed (see e.g. Miller (2002)). In this thesis we suggest another approach: variable selection by genetic algorithms.

The chapter is organized as follows: in the next section we will sketch some common methods for variable selection which are generally used. Section 8.2 presents the structure and the operators of the genetic algorithm for variable selection.

Remark 19. The software of the genetic algorithm for optimization of binary-valued parameters presented in section 8.2 can be downloaded from internet. The address is *http://www.stat.uni-muenchen.de/sfb386/*. A program manual of this software called *GENbin* is available in appendix C.

8.1 Diverse Approaches to Variable Selection

Here we sketch some approaches to variable selection which are commonly used. A detailed presentation about alternative methods can be found in Miller (2002).

In the following we assume that there are p covariates to which the variable selection procedure can be applied. The most usual variable selection procedures are:

- *forward selection*: in this procedure we choose that variable from all variables which yields the smallest information criterion value (e.g. the statistical software package S-Plus uses the AIC information criterion). Let AIC_1 be the corresponding information criterion value. In the next step we choose that variable from the remaining variables for which holds $AIC_2 \leq AIC_1$. We add variables successively and the process continuous until a better (i.e. smaller) value of the AIC cannot be obtained.

- *backward selection*: in contrast to forward selection we start with all p variables (including a constant if there is one). Let AIC_p be the corresponding value of the AIC. That variable is chosen for deletion which yields the smallest value of AIC_{p-1} after deletion. Successively we choose that variable which yields the best (i.e. minimal) value of the AIC and delete it from the model. The process continuous until there is only one variable left, or until a termination condition is achieved.

- *stepwise regression* (or *Efroymson's algorithm* (Miller (2002))): stepwise regression is a combination of forward and backward selection procedures. We assume that -without loss of generality- k variables are in the model. A variable $k + 1$ is included in the model if holds $AIC_{k+1} \leq AIC_k$. After addition of each variable (except the first one) to the set of selected variables we check the AIC if any of the variables previously chosen can be deleted without increasing the AIC. The procedure continuous until addition or deletion yields no improvement of the information criterion.

The selection procedures presented above do not guarantee that the best-fitted subsets are found. Thus it is possible that these approaches lead to bad estimators, especially if some covariates are highly correlated.

In diverse popular statistical software packages all stepwise procedures described above are implemented, e.g. S-Plus, R, BayesX, SAS. Here the software programs use different information criteria for model selection: while S-Plus only use AIC, the programs R and SAS can differ between AIC and BIC. In BayesX we can also choose GCV. In context with the simulation study in chapter 9 we return to the stepwise procedures in S-Plus and R comparing them with the genetic algorithm presented in the next section.

Remark 20. Another approach implemented in many software packages is *subset selection*. In subset selection each possible subset is chosen and the respective value of an information criterion is calculated (e.g in the program R the AIC is chosen). The best subset (i.e. with a minimal value of the applied information criterion) is used as selected model. This approach can only be applied to simple models consisting of a few variables (because of the huge computational costs). For the following simulation studies this procedure is not accurate.

8.2 Structure of a Genetic Algorithm for Variable Selection

The basis of the genetic algorithm for variable selection is an $0-1$ coding of strings. Suppose we have p metrical variables $\mathbf{x}_1, \ldots, \mathbf{x}_p$ and q categorical variables $\mathbf{z}_1, \ldots, \mathbf{z}_q$. Then the coding of the inclusion of metrical variables is given by

$$\delta_j^x = \begin{cases} 1 & if \ variable \ \mathbf{x}_j \ is \ included \\ 0 & else \end{cases} \quad j = 1, \ldots, p,$$

and in case of categorical variables we have

$$\delta_j^z = \begin{cases} 1 & if \ variable \ \mathbf{z}_j \ is \ included \\ 0 & else \end{cases} \quad j = 1, \ldots, q.$$

Interactions are coded in the same way by $\delta_{jk}^{xx}, \delta_{jk}^{zz}, \delta_{jk}^{xz}$ and thus for example δ_{jk}^{xx} is given by

$$\delta_{jk}^{xx} = \begin{cases} 1 & if \ the \ interaction \ between \ \mathbf{x}_j \ and \ \mathbf{x}_k \ is \ included \\ 0 & else \end{cases} \quad,$$

where $j, k = 1, \ldots, p, j \neq k$. It should be noticed that only interactions with $\delta_{jk}^{xx}, j < k$ and $\delta_{jk}^{zz}, j < k$ are used. For interactions between metrical and categorical variables all combinations $\delta_{jk}^{xz}, j = 1, \ldots, p, \ k = 1, \ldots, q$, have to be considered.

It is important to prevent interactions between variables for which no main effects are included. The restriction

$$\delta_{jk}^{xx} \leq \delta_j^x \delta_k^x \tag{8.1}$$

implies that an interaction can be only included if main effects of both variables $\mathbf{x}_j$ and $\mathbf{x}_k$ are included (analogue for categorical variables and their interactions with metrical variables). This relation has always to be checked after application of crossover- and mutation operators to interaction indicators. The indicators may be collected into one string

$$\delta = (\{\delta_j^x\}, \{\delta_j^z\}, \{\delta_{jk}^{xx}\}, \{\delta_{jk}^{zz}\}, \{\delta_{jk}^{xz}\})$$

which only consists of the values 0 and 1. It should be noticed that the length of the components varies, e.g. $\{\delta_{jk}^{zz}\}$ has length $q(q-1)/2$.

Remark 21. If we use interactions between variables we always consider only interactions between two different variables.

Essential components of a genetic algorithm are the diverse operators, i.e crossover- and mutation operator, which we are going to present now.

8.2.1 Adaptive Binary Crossover (ABC)

In context with variable selection or (feature) subset selection diverse crossover operators have been proposed. In most cases they refer to the one-point crossover (see example 3 in section 5.2.2) or its extension to two-point respectively multi-point crossover (Oliveira, Benahmed, Sabourin, Bortolozzi & Suen (2001); Wallet, Marchette, Solka & Wegman (1996); Yang & Honavar (1997)). The authors assume again that the crossover operators chosen have the same influence during the whole application of the genetic algorithm. But as we have seen in section 5.3.1 from the exploitation-exploration-dilemma, the objectives are differently weighted at various times: e.g. at the beginning of the genetic algorithm exploration is more relevant compared with exploitation and vice versa.

To consider the problem of different conflicting objectives during application of a genetic algorithm we propose an adaptive crossover operator and call him *adaptive binary crossover (ABC)*.

Suppose we have two $0-1$ strings with indicator variables $\delta = (\delta_1 \ldots \delta_i \ldots \delta_k)$ and $\bar{\delta} = (\bar{\delta}_1 \ldots \bar{\delta}_i \ldots \bar{\delta}_k)$. A pair of bits $(\delta_i, \bar{\delta}_i)$ of the parent strings swap their places if we have a random number r_i with

$$r_i < p_c \underbrace{(1 - r^{(1-\frac{t}{T})^b})}_{\equiv g(t)}. \tag{8.2}$$

Here $r \in [0,1]$ is a uniform random number equal for all bits of a string, p_c is the crossover probability, t is the current generation number, T is the maximum number of generations and b is a user-dependent system parameter which determines the degree of non-uniformity. Which strings are selected for crossover process is controlled by a similar expression as (8.2).

In contrast to the conventional crossover operators the ABC operator considers the conflicting exploitation and exploration objectives which have different relevance during the application of the genetic algorithm. We can distinguish between two extreme cases (compare also Figure 8.1):

(i) **t small**: in this case the exponent of $g(t)$ is close to zero and hence $g(t)$ is primarily influenced by a suitable choice of the random number r. Figure 8.1 illustrates this context: for a generation number $t = 5$ we receive approximatively a straight line with slope -1. As random number r is uniformly distributed each value $g(t)$ can be (approximatively) accepted with the same probability. If we have chosen $p_c = 1$, diverse strings show many swaps of corresponding bits (if r is small) during crossover process.

By suitable choice of crossover probability p_c the number of swaps between corrsponding bits can be varified. A small value of p_c also decreases the number of swaps between two strings (e.g. a decrease of 0.5 reduces the number of swaps by a half during crossover process).

(ii) **t large**: if the generation number t is large, the exponent of $g(t)$ is close to zero and hence $g(t)$ also yields values close to zero for a wide range of random numbers r. This fact is illustrated in Figure 8.1 for $t = 95$. At the end of the genetic algorithms' application there are only

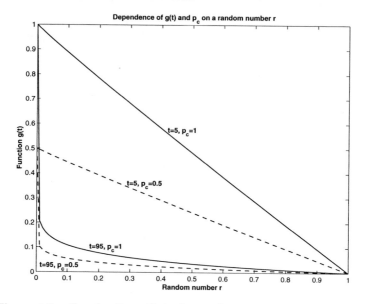

Figure 8.1. *Here function $g(t)$ is shown subject to a uniformly distributed random number r for two sizes of the generation number t and crossover probabilities $p_c = 1$ respectively $p_c = 0.5$. The user-dependent system parameter b is chosen as 1.*

a few swaps between corresponding bits. In addition a decrease of the crossover probability p_c increases the effect.

8.2.2 Adaptive Binary Mutation (ABM)

In context with variable selection or (feature) subset selection the authors generally propose a uniform mutation operator: changing a bit at random (Oliveira, Benahmed, Sabourin, Bortolozzi & Suen (2001); Wallet, Marchette, Solka & Wegman (1996); Yang & Honavar (1997)). But the mutation operator should also consider the relevance of exploitation and exploration at various times during genetic algorithm. Thus we propose an adaptive mutation operator and call it *adaptive binary mutation (ABM)* operator. For each bit of a string we generate a random number r_i and if holds

$$r_i < p_m\bigl(1 - r^{(1-\frac{t}{T})^b}\bigr), \tag{8.3}$$

the bits mutate, i.e. 0 is changed to 1 and vice versa. The idea and functionality of this operator are the same as in the last section about the ABC operator.

8.2.3 Selection Procedure

The genetic algorithm for variable selection also uses the modified selection procedure (modSP) introduced in section 5.3.4. The only difference is the check of available interactions after each crossover respectively mutation step; i.e. as seen in Figure 5.8 of section 5.3.4 we have to introduce a "control step" after step 4 and step 5.

9

Variable Selection in Simulations

In the following simulations we check the performance of variable selection and prediction accuracy of the genetic algorithm presented in chapter 8 and to compare this approach to other methods in literature. Therefore we use additive models consisting of different numbers of variables with linear effect.

To evaluate the performance of an approach to variable selection we have to check the following criteria:

(1) total number of variables which have falsely been classified, i.e. a variable *with* an effect has erroneously been selected as a variable *without* effect and vice versa.

(2) number of variables *with* effect which have erroneously been classified as variables *without* effect;

(3) number of variables *without* effect which have erroneously been classified as variables *with* effect;

(4) the prediction accuracy of the diverse approaches is checked by courtesy of $log(MSE)$.

Beside the choice of default parameters in genetic algorithms also the use of an appropriate information criterion is important. As generally mentioned in chapter 4 the number of relevant variables is reduced during application of a selection procedure. Thus it is possible that an information criterion chosen (with good results at the beginning of the selection procedure) can produce bad results in case of reduced datasets.

9.1 Dependence of Variable Selection on Information Criteria

To check the performance we apply various information criteria to a simulated dataset. We consider an additive model with 5 different functions. Here 3 functions contain variables with linear effect and the other 2 functions have no effect. Hence there are 10 possible interactions between variables x_{ij} and x_{ik}, where $j, k = 1, \ldots, 5, j \neq k$. In this simulation the combinations (x_{i1}, x_{i3}) and (x_{i2}, x_{i3}) only show an effect. Thus the response variable y_i depends on $\mathbf{x}_i$ by

$$
\begin{aligned}
y_i &= f(\mathbf{x}_i) + \epsilon_i \\
&= f_1(x_{i1}) + f_2(x_{i2}) + f_3(x_{i3}) + f_{13}(x_1, x_3) + f_{23}(x_2, x_3) + \epsilon_i \quad (9.1) \\
&= 0.5x_{i1} - 0.4x_{i2} + 0.25x_{i3} + x_{i1}x_{i3} + x_{i2}x_{i3} + \epsilon_i
\end{aligned}
$$

with $i = 1, \ldots, n$. We simulate 200 datasets, each one consisting of $n = 200$ observations, independently and uniformly distributed. The noise has standard deviation $\sigma_1 = 0.1$.

As default parameters of the genetic algorithm are used: population size $(popsize) = 48$ strings, crossover probability $p_c = 1$, mutation probability $p_m = 1$, deletion of $u = 60$ percent of the worst strings, selection of $r = 30$ and $s = 18$ strings, $\nu = 0.5, T = 1000$ and $b = 1$.

For the information criteria presented in chapter 4, Figure 9.1 shows the number respectively portion of datasets with a total number of $0, 1, 2 \ldots$ misclassified variables. Here e.g. the black coloured part of the bars yields the number (respectively portion) of datasets with no incorrectly classified variable, i.e. each variable with effect is classified as variable with effect and vice versa. We realize that the BIC has by far the lowest misclassification rate: in more than 90% of the datasets all variables are correctly classified. All the other information criteria only show approximately $40 - 50\%$ datasets which are completely correctly classified. Furthermore the BIC yields datasets with at maximum two incorrectly classified variables. In contrast the other information criteria generate datasets with at maximum six incorrectly classified variables.

Without any plot we have to notice that in each dataset of this simulation study all information criteria correctly classify the variables with effect.

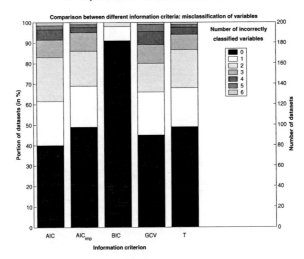

Figure 9.1. *The simulation results of the additive model in (9.1) with $\sigma_1 = 0.1$ for diverse information criteria are shown here. The panel yields the number (respectively portion) of datasets with incorrectly specified variables.*

The top panel of Figure 9.2 analyse the prediction accuracy of the diverse information criteria by plotting $log(MSE)$. Here BIC yields the best (i.e. the lowest values for $log(MSE)$) results compared with the other criteria. The other plot of Figure 9.2 shows the number of iterations up to the incidence of the result used in the evaluation (i.e. the incidence of the result is connected with the best fitness value in the simulation of a dataset). Here we receive a significantly lower iteration number of BIC compared with all other information criteria used. The results presented above suggest the application of BIC in the following simulation studies.

Remark 22. Compared with the number of possible variables the simulation above uses a large number of observations (i.e. $n = 200 \gg 15$ *variables*). Hence the expression $tr(\mathbf{H})/n$ in Figure 4.1 takes small values. In these cases BIC with its stronger penalization yields better results in variable selection. If we have datasets containing only a few observations as well as a large number of possible variables (e.g. analysis of gene expression data) the penalization of BIC is too rigid. Thus other criteria (e.g. AIC or improved AIC) yield better results. Depending from the respective dataset used in the further course of this thesis we often switch between BIC, AIC and improved AIC.

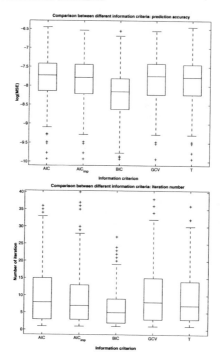

Figure 9.2. *The simulation results of the additive model in (9.1) with $\sigma_1 = 0.1$ for diverse information criteria are shown here. The top panel checks the prediction accuracy by calculating $log(MSE)$. The panel below yields the number of iterations up to the incidence of the result used in the evaluation.*

9.2 Variable Selection in Additive Models with different Complexity

To compare the performance of the genetic algorithm for variable selection with other approaches in literature we have chosen two alternative methods:

- The software package S-Plus offers a possibility of variable selection based on stepwise procedures. First one calculates the AIC for an initial model. Then each covariate can be dropped or integrated in the model as a linear term. Starting with the initial model the implemented function **step** successively calculates the AIC for all alternative models. If a current model yields a better AIC-value the previous model is replaced. The user can choose between forward- and backward-selection respectively stepwise regression procedures.

- Moreover the statistic software package R offers an approach to variable selection. The function stepAIC implemented in the package MASS chooses a model by the AIC in a stepwise algorithm. This procedure is comparable with the step-function in S-Plus. Again the user has the possibility to choose between forward- and backward-selection respectively stepwise regression procedures. Beside AIC the user can also choose BIC as criterion. In the simulations we have applied the BIC.

The dataset of the following simulation study bases on the true function (9.1). Beside the case of $\sigma_1 = 0.1$ we also analyse the case with noise $\sigma_2 = 0.2$.

Figures 9.3 and 9.4 yield the results of criteria (1) − (3) for the function (9.1) with $\sigma_1 = 0.1$ respectively $\sigma_2 = 0.2$. From these Figures we can draw the following conclusions:

- In case of $\sigma_1 = 0.1$ the genetic algorithm yields significantly better results compared with the stepwise regression procedures in S-Plus and R. The bar graphs in the first plot show that in approximately 90% of the datasets the number of incorrectly specified variables is zero. In the other approaches, however, approximately 15% datasets have completely correctly specified variables, only. Furthermore the genetic algorithm has datasets with at most two incorrectly specified variables. In approximately 60% datasets the other approaches only have at maximum two incorrectly classified variables. Besides also datasets with up to seven incorrectly classified variables are possible.

The other two plots illustrate, whether the variables with or without effect are correctly specified. The second plot shows that in almost all datasets the variables with effect are correctly detected by the three software programs. Thus we conclude that errors appear in variables without effect, i.e. these variables are erroneously classified as variables with effect. This fact can be seen in the plot below: here in approximately 9% of the datasets the genetic algorithm erroneously selects more variables with effect as actually included in the additive model (9.1). For the other approaches the portion has larger values.

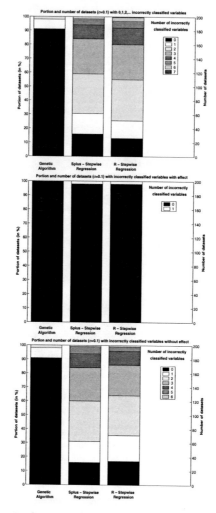

Figure 9.3. *The simulation results of the additive model in (9.1) with $\sigma_1 = 0.1$ are shown here. The first plot yields the number (respectively portion) of datasets with incorrectly specified variables, while the second one checks the errors, restricted to variables with effect. The plot below yields comparable results, restricted to variables without effect.*

- In case of $\sigma_2 = 0.2$ the genetic algorithm yields similar results compared with $\sigma_1 = 0.1$. The bar graphs of the other approaches, however, show a larger number of datasets with incorrectly classified variables: e.g.

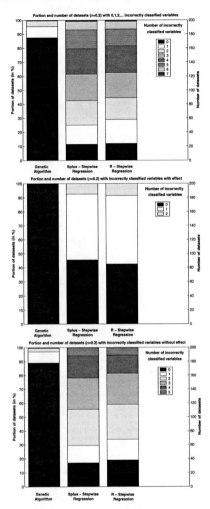

Figure 9.4. *The simulation results of the additive model in (9.1) with $\sigma_2 = 0.2$ are shown here. The first plot yields the number (respectively portion) of datasets with incorrectly specified variables, while the second one checks the errors, restricted to variables with effect. The plot below yields comparable results, restricted to variables without effect.*

approximately 45% datasets have at most two erroneously classified variables (compared with 60% in case of $\sigma_1 = 0.1$).

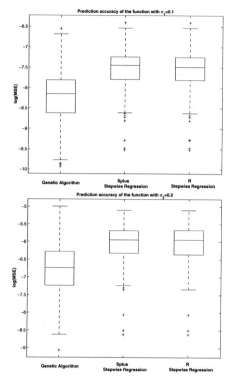

Figure 9.5. *Here prediction accuracy for the genetic algorithm and the stepwise regression procedures in S-Plus and R are shown. The plot above yields the case of the additive model (9.1) with $\sigma_1 = 0.1$. The plot below yields the other case for $\sigma_2 = 0.2$.*

The distribution of errors shows a significant difference: in contrast to the case $\sigma_1 = 0.1$ more than a half of the datasets in S-Plus and R show misclassifications of variables with effect, i.e. in some datasets variables with effect are erroneously classified as variables without effect. Application of the genetic algorithm does not show this problem: in all datasets variables with effect are detected. Thus all errors in variable selection of the genetic algorithm only appear in variables without effect.

Besides the number of misclassifications in a variable selection also prediction accuracy plays an important role. The boxplots in Figure 9.5 show the prediction accuracy of the diverse approaches for the additive model

(9.1) with $\sigma_1 = 0.1$ and $\sigma_2 = 0.2$. In both cases the genetic algorithm yields significantly better estimations (i.e. lower $log(MSE)$-values) compared with the stepwise regression procedures in S-Plus and R. Thus a lower number of variable misclassifications in the genetic algorithm also results in a higher prediction accuracy.

The next simulation study analyses the performance of the genetic algorithm in case of a significant increase of variables. For that purpose we consider an additive model consisting of 10 different functions. Here only 5 functions contain variables with a linear effect. Hence there are 45 interactions possible between variables x_{ij} and x_{ik}, where $j, k = 1, \ldots, 10, j \neq k$. In this simulation only 6 combinations show any effect. Again we simulate 200 datasets, each one consisting of $n = 500$ observations independent and uniformly distributed. The noise has standard deviation $\sigma = 0.1$.

The top panel of Figure 9.6 shows the number of variables which have falsely been classified. For more clarity the datasets are divided into four classes, in fact datasets with $0 - 2$ errors, $3 - 5$ errors and $6 - 8$ errors respectively to datasets with more than 8 errors. The genetic algorithm yields significantly better results compared with the stepwise regression procedures in S-Plus and R. The bar graphs of the genetic algorithm show that in 44% of the datasets the number of incorrectly specified variables is between 0 and 2. The other approaches have no (S-Plus) respectively only a few (R) datasets consisting $0 - 2$ errors. S-Plus and R also yield 5.5% respectively 3% datasets with at maximum 5 errors. In contrast the genetic algorithm yields 95% datasets with at most 5 errors. The biggest part of datasets (70%) is classified with more than 8 errors by S-Plus and R. This problem of large misclassification is yielded by the genetic algorithm in a few datasets only (6%).

The panel below in Figure 9.6 illustrates the number of datasets with errors in variables with effect (i.e. erroneously classified as variables without effect). Here we realize that the genetic algorithm classifies the variables with effect correctly in approximately 20% datasets. This number is slightly lower compared with the stepwise procedures of S-Plus and R. Application of the genetic algorithm with the improved AIC information criterion would increase the set of datasets with correctly classified variables with effect to approximately 53%.

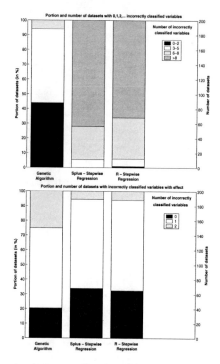

Figure 9.6. *The simulation results of the large additive model with σ = 0.1 are shown here. The first plot yields the number (respectively portion) of datasets with incorrectly specified variables, while the second one checks the errors restricted to variables with effect.*

All the three approaches show a relatively low error rate in detecting variables with effect (at most 2). Thus we conclude that errors principally appear in variables without effect, i.e. these variables are erroneously classified as variables with effect.

Finally Figure 9.7 shows prediction accuracy of the diverse approaches influenced by different performance in variable selection. It is obvious that the genetic algorithm yields significantly better estimations (i.e. lower $log(MSE)$-values) compared with the stepwise regression procedures in S-Plus and R. Hence this simulation study results in a higher prediction accuracy in case of lower variable misclassifications, too.

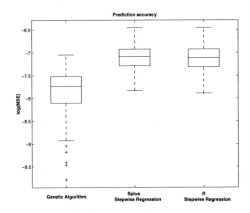

Figure 9.7. *Here prediction accuracy for the genetic algorithm and the stepwise regression procedures in S-Plus and R for the large additive model is shown.*

Variable Selection of Gene Expression Data

An interesting application of algorithms for variable selection is the analysis of gene expression data. Here questions respective responsibility of special genes can be treated by means of microarray technology, if, for example, the user is interested in the knowledge of which genes are involved in the creation of diseases (e.g. cancer). Usually such studies consist of many thousands of genes but only of a few samples. For a detailed presentation of microarray technology as well as approaches to the extraction of gene expression data we refer e.g. to Hamadeh & Afshari (2000).

In context to this chapter we analyse colon data containing expression levels of 2000 genes $x_{ij}, j = 1, \ldots, 2000$, and 62 samples $y_i, i = 1, \ldots, 62$. Here 40 samples descend from patients with tumor tissue and 20 samples descend from patients with normal tissue. For further details see Alon, Barkai, Notterman, Gish, Ybarra, Mack & Levine (1999). The dataset can be downloaded from *http://microarray.princeton.edu/oncology*.

By application of the genetic algorithm GENbin from the set comprising 2000 genes we search a subset relevant for observations. The dataset is divided into a training dataset with 41 samples and a test dataset with 21 samples. Here we have 200 replications of a training dataset with randomly chosen observations. The remaining samples form the test datasets.

10.1 Description of the Approaches used

In this section we briefly describe the modifications of the genetic algorithm GENbin for application to the gene expression dataset. Furthermore we present some alternative approaches used for comparison.

10.1.1 Modifications of the Genetic Algorithm GENbin

The genetic algorithm for variable selection used so far can be taken with some small modifications.

To generate the initial population we first calculate the fitness values for all subsets containing only one gene. That means for the colon dataset used that we have 2000 fitness values (because of the 2000 different genes). In the initial population we use the *popsize* strings with the best fitness values (in our case the initial population consists of *popsize* = 48 strings).

As fitness function we use the information criteria AIC and BIC. But here the criteria are based on binomial distributed observations. As modelling procedure we apply the logistic regression.

As mentioned in section 2.2 the probability $\pi(\mathbf{x}_i)$ to get an event $A(y_i = 1)$ for given variables $\mathbf{x}_i$ can be received by application of the logistic regression. Assuming a logistic model

$$\pi(\mathbf{x}_i) = \frac{1}{1 + e^{-(\beta_0 + \sum_{j=1}^{p} \beta_j x_{ij})}} \quad , \tag{10.1}$$

we get estimators for the probabilities. Here the unknown parameters $\beta_0, \beta_1, \ldots, \beta_p$ have to be estimated appropriately. Appropriate estimators can be received for independent identical Bernoulli distributed observations $(y_i \sim \mathcal{B}(1, \pi_i), i = 1, \ldots, n)$ by maximization of the likelihood function. Here the likelihood function takes the form

$$L = \prod_{i=1}^{n} \pi_i^{y_i} (1 - \pi_i)^{1-y_i} \quad . \tag{10.2}$$

Logarithmizing of formula (10.2) we receive the respective log-likelihood function

$$l = \sum_{i=1}^{n} y_i \, log(\pi_i) + \sum_{i=1}^{n} (1 - y_i) \, log(1 - \pi_i) \quad . \tag{10.3}$$

Inserting of (10.1) into the log-likelihood function and maximization yields the appropriate parameter estimator (for futher details see e.g. Fahrmeir & Tutz (2001)).

As mentioned in chapter 4 it is necessary in variable selection to evaluate prediction accuracy and model complexity. In context with this chapter we use AIC

$$AIC = -2l + 2q$$

respectively BIC

$$BIC = -2l + q\,log(q)$$

as fitness functions, where $q = p + 1$ is the number of paramters in the model which have to be estimated. In search of an appropriate model AIC (respectively BIC) has to be minimized.

Because of the small number of samples as well as the large number of genes in the dataset used, selection of too many genes leads to numerical instabilities during estimation. To prevent these problems we restrict the number of selected genes to maximal 10 genes (i.e. each string contains 10 genes at maximum).

As default parameters of the genetic algorithm are used: population size ($popsize$) = 48 strings, crossover probability $p_c = 1$, mutation probability $p_m = 1$, deletion of $u = 60$ percent of the worst strings, selection of $r = 38$ and $s = 10$ strings, $\nu = 0.5$, $T = 1000$ and $b = 1$.

10.1.2 Genetic Algorithm including "Response-Smoothing-Estimation"

Also in this section we assume a logistic model and use the modified genetic algorithm of section 10.1.1. However, to improve the results the parameters are estimated by a shrinkage approach called "Response-Smoothing-Estimation" (Tutz (2003)). Here the reponse variables y_i are appropriately transformed.

The approach is based on the following idea: for each observation $(y_i, \mathbf{x}_i)$, $i = 1, \ldots, n$, a pseudo-observation $(1 - y_i, \mathbf{x}_i), i = n + 1, \ldots, 2n$, is introduced. Thus the dataset consists of $2n$ "observations". Furthermore we introduce weights

$$w_i = \begin{cases} 1 - \alpha_i & if \quad i \leq n \\ \alpha_i & if \quad i > n \end{cases},$$

where $\alpha_i \in [0, 0.5]$. By the weights the response variables y_i can be transformed to

$$y_i^{(t)} = y_i + (-1)^{y_i}\alpha_i \quad . \tag{10.4}$$

We realize that the response variables $y_i = 0$ are transformed to larger values $y_i^{(t)} = \alpha_i$ while the response variables $y_i = 1$ are transformed to smaller values $y_i^{(t)} = 1 - \alpha_i$. Hence the response variables are shifted in direction of the value 0.5 (which is the mean of the logistic distribution, compare Figure 2.2). With the new transformed response variables $y_i^{(t)}$ we execute a maximum-likelihood estimation as described e.g. in Fahrmeir & Tutz (2001).

The transformation of the data can be motivated by the fact that the probabilities π_i of a binary regression model have to be estimated. However $y_i \in \{0, 1\}$ can be assumed as extremely unrealistic probabilities. Thus by transformation the y_i are approximated to the true probabilities.

For further details as well as a suggestion for the choice of the tuning parameter α_i which are based on cross-validation we refer to Tutz (2003).

With an approach involving diagnostic measures (Leitenstorfer (2003)) we receive a formula for the calculations of α_i:

$$\alpha_i = \nu_i \left[h_{ii} + \frac{\chi_i^2}{\chi^2} \right]^c , \qquad (10.5)$$

where

$$\nu_i = \begin{cases} \bar{y} & if \;\; y_i = 0 \\ 1 - \bar{y} & if \;\; y_i = 1 \end{cases} , \bar{y} = \frac{1}{n} \sum_{i=1}^{n} y_i .$$

and $c \in [0, \infty)$. The elements h_{ii} belong to the generalized hat matrix $\mathbf{H} = \mathbf{W}^{1/2} \mathbf{X} (\mathbf{X}^T \mathbf{W} \mathbf{X})^{-1} \mathbf{X}^T \mathbf{W}^{1/2}$ with weight matrix $\mathbf{W}$ (compare e.g. Fahrmeir & Tutz (2001)) and $\chi^2 = \sum_{i=1}^{n} \chi_i^2$ is the χ^2-goodness-of-fit statistic with components

$$\chi_i = \frac{y_i - \pi_i}{\sqrt{\pi_i(1 - \pi_i)}} .$$

The expression $h_{ii} + (\chi_i^2/\chi^2)$ in formula (10.5) is a measure to check the position in the design space as well as the prediction accuracy by the model of each observation. Large values of (10.5) belong to observations which are outliers (large value of h_{ii}) or badly fitted by the model (large value of the second term).

To control the size of the shrinkage effect the user has to choose the tuning constant c. For $c \in [0, 1)$ we receive a larger α_i and hence a stronger shrinkage effect. In case of $c = 0$ we get a total smoothing. If the user chooses $c \in (1, \infty)$ the α_i takes smaller values and thus a decreasing shrinkage effect (i.e. the estimations are closer to the maximum-likelihood estimations).

In case of the calculations with the colon dataset we choose $c = 0.3$ and $c = 1.0$.

10.1.3 Nearest-Neighbour Method

The k-nearest-neighbour method requires no model to be fit and works in the following way: given a point $\mathbf{x}_i^\star, i = 1, \ldots, n$, in the training dataset we choose the k nearest neighbour by the Euklidian distance

$$d_i = \|\mathbf{x}_i - \mathbf{x}_i^\star\| \ .$$

An estimator $\hat{y}_i$ can be received by

$$\hat{y}_i = \frac{1}{k} \sum_{\mathbf{x}_i \in N_k(\mathbf{x}_i^\star)} y_i \ ,$$

where $N_k(\mathbf{x}_i^\star)$ is the neighbourhood of $\mathbf{x}_i^\star$ defined by the k closest points $\mathbf{x}_i$ in the training sample. If the estimator $\hat{y}_i$ takes a value > 0.5 we assign $\hat{y}_i$ the value 1 (i.e. the patient has tumor). Otherwise ($\hat{y}_i \leq 0.5$) the estimator is assigned the value 1 (i.e. the patient has no tumor). By comparison with the true observations y_i the number of errors can be determined. And this error rate has to be minimized.

For application of the k-nearest-neighbour method to the colon dataset we choose $k = 1$ and $k = 5$.

10.1.4 Discrete AdaBoost

The motivation for the discrete AdaBoost procedure (Friedman, Hastie & Tibshirani (2000)) was to combine the outputs of many "weak" classifiers to produce a powerful "comitee". The algorithm works in the following way:

(1) Each observation of the training dataset is provided with an initial weight $w_i = 1/m, i = 1, \ldots, m$.

(2) For $t = 1 : M$

    (a) Use a classifier $G_t(x)$ (e.g. CART, compare section 10.1.5) and fit the classifier to the training data which use the weights w_i. The classifier $G_t(x)$ produces a prediction taking one of the two variables $\{-1, +1\}$, i.e. each element of the training sample is assigned a prediction $\in \{-1, +1\}$.

(b) The resulting weight error rate is computed by

$$err_t = \frac{\sum_{i=1}^{m} w_i I(y_i \neq G_t(x_i))}{\sum_{i=1}^{m} w_i} \quad ,$$

i.e. in case that the true observation y_i and the prediction (produced by the classifier $G_t(x)$) are different, the error rate increases by a weighted amount.

(c) Compute $\alpha_t = log((1 - err_t)/err_t)$. α_t weights the influence of the used classifier.

(d) Set

$$w_i \equiv w_i \cdot e^{\alpha_t I(y_i \neq G_t(x_i))} \qquad , i = 1, \ldots, m \quad .$$

The individual weights of each observation is updated for the next iteration. We realize that the misclassified observations are stronger weighted by the term $exp(\alpha_t)$. The objective is that the next classifier $G_{t+1}(x)$ stronger influences the larger weighted observations.

(3) The predictions from all classifiers $G_t(x), t = 1, \ldots, M$, are combined through a weighted sum

$$G(x) = sign \left[\sum_{t=1}^{M} \alpha_t G_t(x) \right] \quad .$$

In the colon dataset we have chosen the CART procedure (compare section 10.1.5) as classifier. The number M of iterations has the size 50.

10.1.5 CART

Classification and regression trees (in short CART) have been developed by Breiman, Friedman, Olshen & Stone (1984). The idea is that the predictor space is successively divided and the resulting splits have to be heterogeneous as much as possible in respect of variable y. Otherwise the values have to be homogeneous within a split. For example let only one metrical variable x be given; thus we search for a cutting point c with the property: the split in sets $A_1 = \{x : x \leq c\}$ respective $A_2 = \{x : x > c\}$ lead to similar values within the sets but different values between A_1 and A_2 (i.e. $y = 0$ in A_1 and $y = 1$ in A_2 or vice versa).

In case of the colon dataset in each iteration the optimal split is searched for each variable. The variable which yields the best split is selected. The maximal number of splits is determined as 4, i.e. we have 5 selected variables.

10.2 Results of the Colon Dataset

For the diverse approaches described in section 10.1 the following Table 10.1

Software program	Average misclassification rate	Standard deviation
Genetic algorithm (AIC)	0.2348	0.0928
Genetic algorithm (BIC)	0.2264	0.0916
Genetic algorithm with Resp.-Smooth.-Estim. $c = 0.3$ (AIC)	0.2281	0.0861
Genetic algorithm with Resp.-Smooth.-Estim. $c = 1.0$ (AIC)	0.2317	0.0894
Genetic algorithm with Resp.-Smooth.-Estim. $c = 1.0$ (BIC)	0.2224	0.0888
Discrete AdaBoost	0.1920	0.0720
CART	0.2930	0.0890
1-Nearest neighbour	0.2520	0.0860
5-Nearest neighbour	0.2140	0.0890

Table 10.1. *For the software programs described in section 10.1 the average misclassification rate and the standard deviation of the misclassification rates for 200 replications of the test dataset are shown. Beside discrete AdaBoost, CART and Nearest neighbour (with $k = 1$ respectively $k = 5$) approaches the genetic algorithm including AIC and BIC is observed. Furthermore the modified genetic algorithm with "Response-Smoothing-Estimation" and different parameter c is used: $c \in \{0.3, 1.0\}$ in case of AIC and $c = 1.0$ for the genetic algorithm including BIC.*

yields the average misclassification rate respectively the standard deviation of the misclassification rates for 200 replications of the test dataset. The average misclassification rate is defined as

$$\overline{err} = \frac{1}{N} \sum_{l=1}^{N} \left[\frac{1}{m} \sum_{i=1}^{m} I\left(\hat{y}_i, y_i\right) \right] \quad ,$$

where N is the number of replications of the test dataset (here $N = 200$), m is the number of samples in the test dataset ($m = 21$) and

$$I(\hat{y}_i, y_i) = \begin{cases} 0 & if \quad \hat{y}_i = y_i \\ 1 & if \quad \hat{y}_i \neq y_i \end{cases}.$$

Beside discrete AdaBoost, CART and nearest neighbour methods ($k = 1, 5$) the results of the genetic algorithm including AIC respectively BIC are also presented in Table 10.1. Furthermore we analyse the genetic algorithm of section 10.1.2 which is characterized by the "Response-Smoothing-Estimation" approach.

Compared with the other approaches of Table 10.1 we realize that average misclassification rate and standard deviation take the smallest values in case of discrete AdaBoost. Except for discrete AdaBoost all other methods show comparable values for standard deviation: the values differ between 0.0860 and 0.0928. The nearest neighbour approach with $k = 5$ and the diverse genetic algorithms yield similar results for the average misclassification rate. Significantly worse results are shown by the nearest neighbour method with $k = 1$ and the CART approach.

Furtermore it is obvious that the genetic algorithm including the "Response-Smoothing-Estimation" leads to slightly better results compared with the genetic algorithm without modification, i.e. we have a smaller average misclassification rate respectively standard deviation.

With respect to the different information criteria we detect smaller average misclassification rates as well as standard deviations for genetic algorithms with BIC. However the differences are relatively slight: e.g. the average misclassification rate of the two genetic algorithms including AIC differ from each other in approximately 3% and in case of standard deviation in approximately 7%.

In addition to the results in the table above Figure 10.1 yields portion respectively number of datasets with $0, 1, 2, \ldots$ incorrectly classified observations. Here the number of errors has been divided into four classes, in fact $0-2$, $3-5$, $6-8$, and $9-11$ misclassified observations per test dataset. Corresponding with Table 10.1 also Figure 10.1 shows that AdaBoost has fewer incorrectly classified observations in a test dataset compared with all other approaches. Except CART and nearest neighbour method with $k = 1$ the remaining approaches (i.e. diverse genetic algorithms as well as nearest neighbour method with $k = 5$) show similar results. However CART and the nearest neighbour method with $k = 1$ yield significantly worse results.

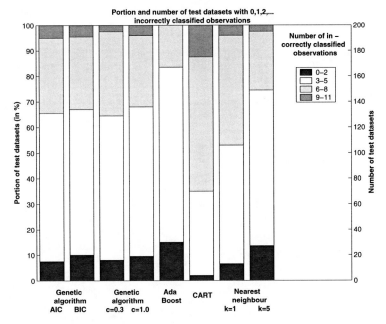

Figure 10.1. *For the different software approaches of section 10.1 this panel shows portion respectively number of test datasets with 0, 1, 2, . . . incorrectly classified observations. The results for the genetic algorithm including "Response-Smoothing-Estimation" are shown for the two best cases c = 0.3 (AIC) and c = 1.0 (BIC).*

Another important question is the frequency of occurrence of diverse genes. Especially in datasets consisting of a large number of genes (in our case we have 2000 genes) the geneticist is interested in an appropriate restriction to only a few relevant genes (which can further be analysed). It is useful to check their occurence in a dataset because frequently occurring genes usually have a significant influence in the dataset.

In Figure 10.2 the frequency of occurence is shown in dependence on the gene number. The results are calculated by the genetic algorithm of section 10.1.1 including AIC (top panel) respectively BIC (panel below). By application of the genetic algorithm to the 200 datasets in each case we get a combination of selected genes. We realize that some genes over and over occur in many training datasets. Furthermore the genetic algorithm ignores most genes of the dataset or chooses them only one time.

Because of the stronger penalization of the BIC usually a smaller number of relevant genes is choosen from the 200 training datasets (i.e. for example many genes chosen by the genetic algorithm with AIC are not included in case of BIC). Furthermore the frequency of occurrence of a few distinguished genes is larger than in case of the genetic algorithm with AIC. However similar genes are preferred by the genetic algorithm with both information criteria.

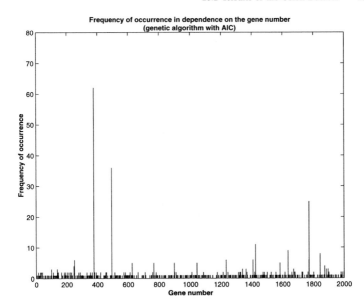

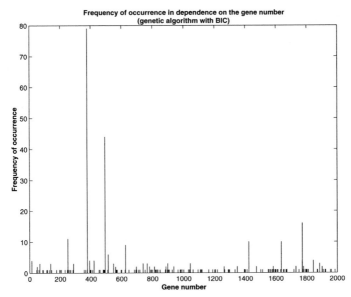

Figure 10.2. *The panels show the frequency of occurrence in dependence on the gene number for the colon dataset. The top panel yields the results for the genetic algorithm including AIC, the panel below uses the same algorithm including BIC.*

Part IV

Simultaneous Selection of Variables and
Smoothing Parameters with Genetic
Algorithms

The Concept of Simultaneous Selection of Variables and Parameters

As mentioned in chapter 8 the problem of finding an appropriate variable selection is of substantial interest, because in many statistical applications (e.g. analysis of gene expression data) there are large sets of explanatory variables which contain many redundant or irrelevant variables. Hence retaining this class of variables leads to inaccurate estimations. On the other side the user expects a significant improvement of the functions' estimation by an appropriate variable selection.

To receive an appropriate estimation of the diverse terms of an additive model, we again choose the approach of using a large number of basis functions with penalization of the coefficients (compare chapter 3). The danger of overfitting, resulting in wiggly estimated curves, is avoided by introducing a penalty term characterized by a smoothing parameter λ. The smoothing parameter controls the influence of the penalty term and hence the smoothness of the estimated function. A large parameter value yields smooth estimates (e.g. $\lambda \to \infty$ leads to a linear estimator). In contrast, a small parameter value yields wiggly estimated curves. To prevent over- respectively underfitting of data an accurate choice of the smoothing parameter is essential.

Many software packages have separate tools for variable selection and smoothing parameter choice which are applied successively. To our knowledge no common statistical software program contains a complete automatical procedure which simultaneously yields variables and smoothing parameters without any additional restrictions. Here we propose simultaneous selection of variables and smoothing parameters by application of genetic algorithms. First approaches in this direction are published in Krause & Tutz (2004).

Remark 23. Even if we are restricted in this thesis to simultaneous selection of variables and smoothing parameters the problem (respectively the corresponding genetic algorithm) can be generalized to any selection of variables and parameters. Hence the choice of smoothing parameters is only one important application of this selection procedure.

The chapter is organized as follows: the next section yields diverse operators and tools of the genetic algorithm for simultaneous selection of variables and smoothing parameters. In section 11.2 the diverse tools for selection of variables and smoothing parameters are combined and integrated in the simultaneous genetic algorithm.

11.1 The Elements of the Simultaneous Genetic Algorithm

In contrast to the genetic algorithms presented above, in the present problem the strings of the population are a combination of a $0-1$ string δ coding the presence of the diverse variables and a real-valued string λ of smoothing paramters.

Suppose we have p metrical variables $\mathbf{x}_1, \ldots, \mathbf{x}_p$ and q categorical variables $\mathbf{z}_1, \ldots, \mathbf{z}_q$. Then the coding of the inclusion of metrical variables is given by

$$\delta_j^x = \begin{cases} 1 & \text{if variable } \mathbf{x}_j \text{ is included} \\ 0 & \text{else} \end{cases} \qquad j = 1, \ldots, p,$$

and in case of categorical variables we have

$$\delta_j^z = \begin{cases} 1 & \text{if variable } \mathbf{z}_j \text{ is included} \\ 0 & \text{else} \end{cases} \qquad j = 1, \ldots, q.$$

Interactions are coded in the same way by $\delta_{jk}^{xx}, \delta_{jk}^{zz}, \delta_{jk}^{xz}$ and thus for example δ_{jk}^{xx} is given by

$$\delta_{jk}^{xx} = \begin{cases} 1 & \text{if the interaction between } \mathbf{x}_j \text{ and } \mathbf{x}_k \text{ is included} \\ 0 & \text{else} \end{cases},$$

where $j, k = 1, \ldots, p, j \neq k$. It should be noted that only interactions with $\delta_{jk}^{xx}, j < k$ and $\delta_{jk}^{zz}, j < k$ are used. For interaction between metrical and categorical variables all combinations $\delta_{jk}^{xz}, j = 1, \ldots, p, \; k = 1, \ldots, q$, have to be considered. The indicators may be collected into one string

$$\delta = (\{\delta_j^x\}, \{\delta_j^z\}, \{\delta_{jk}^{xx}\}, \{\delta_{jk}^{zz}\}, \{\delta_{jk}^{xz}\})$$

which only consists of the values 0 and 1.

For the sake of interpretability hierarchical models are preferred. Thus the model term is restricted by

$$\delta_{jk}^{xx} \leq \delta_j^x \delta_k^x \tag{11.1}$$

which implies that an interaction can only be included if both variables $\mathbf{x}_j$ and $\mathbf{x}_k$ are included. The same is postulated for categorical variables and their interactions with metrical variables.

Each indicator string δ in the population is connected to a smoothing parameter string

$$\lambda = (\{\lambda_j^x\}, \{\lambda_{jk}^{xx}\}, \{\lambda_{jk}^{xz}\}) \; ,$$

which contains the smoothing parameters for the corresponding variables. Here $\{\lambda_j^x\}, j = 1, \ldots, p$, describes the set of smoothing parameters belonging to the metrical variables (without interactions) $\mathbf{x}_1, \ldots, \mathbf{x}_p$. In case of interactions we have similar expressions $\{\lambda_{jk}^{xx}\}$ and $\{\lambda_{jk}^{xz}\}$.

In contrast to δ, λ only contains three elements. The reason is that categorical variables as well as their interactions do not have any smoothing parameters which have to be optimized (compare section 3.3). In the following the combined string is denoted by (δ, λ).

To construct a genetic algorithm for simultaneous selection of variables and smoothing parameters, the following operators, presented above can be directly used:

- *adaptive binary mutation* (*ABM*, section 8.2.2) as mutation operator of the variable strings;

- *non-uniform mutation* (section 5.3.3) as mutation operator for the parameter strings;

- *adaptive binary crossover* (*ABC*, section 8.2.1) as crossover operator for the variable strings.

Only the *improved arithmetical crossover* operator (*IAC*) of section 5.3.2 for the crossover process of the parameter strings has to be adapted to the

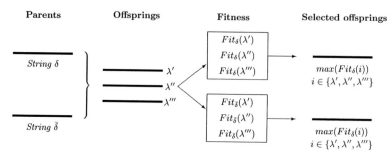

Figure 11.1. *Crossover of two randomly chosen parent strings δ and $\bar{\delta}$ with different combinations of variables yield three offsprings $(\lambda', \lambda'', \lambda''')$. Depending on the variables which are used by parents the fitness of the offsprings is calculated. The two offsprings with the maximal fitness values replace the parent strings.*

actual problem. This operator type has been constructed for the case that the whole population of strings consists of the same number of variables and hence the fitness only depends on the different smoothing parameter values. This does not hold in simultaneous selection of variables and smoothing parameters. Two strings δ and $\bar{\delta}$ containing different combinations of variables generally yield different fitness values. Now we modify the IAC operator in such a way that for each of the two parent strings taking part in the crossover process, the fitness of the three offsprings is separately calculated. Then the best offspring of each group replaces the appropriate parent string (Figure 11.1). We denote the modified IAC operator as *modIAC*.

11.2 The Simultaneous Genetic Algorithm

To receive a genetic algorithm for simultaneous selection of variables and smoothing parameters the diverse tools and operators of section 11.1 have to be appropriately combined.

Generally an indicator string δ contains a mix of elements with values 0 and 1, i.e. only some variables (expressed by value 1) are contained in the string. Hence for the smoothing parameter string λ we only have smoothing parameters in case that the respective element of the indicator string takes value 1. If an element of the indicator string has the value

0 the respective smoothing parameter value is chosen as 0. In case of a metrical indicator variable without interaction that means

$$\lambda_j^x \in \begin{cases} \mathbb{R}_0^+ & if\ \delta_j^x = 1 \\ \{0\} & if\ \delta_j^x = 0 \end{cases} \quad j = 1, \ldots, p.$$

But in the genetic algorithm presented here in case of $\delta_j^x = 0$ the smoothing parameter λ_j^x is not chosen as 0 but retains its actual smoothing parameter. This fact has the following advantage: if the indicator variable again changes from 0 to 1 (e.g. by mutation) the respective smoothing parameter has not been randomly selected which generally leads to results far away from any optima. Instead we assume that the actual smoothing parameter which is already determined in former iterations is in the surroundings of an optimum. Further application of the genetic algorithm tries to find more fit offsprings which are more close to the optimum (exploitation). Thus we do not permanently have to explore the whole search space for better solutions.

In the simultaneous genetic algorithm the mutation operators for selection of variables respectively smoothing parameters run dependently. In the mutation procedure at first randomly chosen elements of the indicator string δ are mutated by use of the adaptive binary mutation (ABM). Afterwards the non-uniform mutation operator is applied to the elements of the respective string λ which belong to the mutated elements of δ. The other smoothing parameters remain unchanged.

Different from mutation the crossover operators (ABC and IAC) run simultaneously but independently from each other. Own simulation trials have shown that it is favourable to use different crossover rates for variables and parameters. Here the crossover rate for the variables is lower than that for smoothing parameters, i.e. the number of crossover processes for variables is lower.

As selection procedure we can use the *modified selection procedure (modSP)* presented in section 5.3.4, but slightly adapted to the actual optimization problem. Also the selection procedure has a link between variables and smoothing parameter selection. In case of a chosen indicator string δ also the respective smoothing parameter string λ is selected, i.e. the couples (δ, λ) remain unchanged. Then this adapted selection procedure including crossover- and mutation operators consists of 9 steps and is illustrated in Figure 11.2:

Step 1: In iteration step t population $P(t)$ of $m = r + s$ strings (δ, λ) is generated by selecting from the previous population. Then the worst u percent strings of $P(t)$.

Step 2: From the remaining strings of step 1 randomly r strings (δ, λ) are selected. These strings do not necessarily have to be distinct.

Step 3: From the remaining strings of step 1 randomly select $s = m - r$ parent strings (δ, λ) are selected. These have not to be distinct from the r selected strings in step 2.

Step 4: If identical strings are in the population (i.e. all genes of the strings are identical) the copies will be mutated by using the ABM operator on the indicator strings δ. How many genes of a string are randomly selected and mutated is controlled by a random number (at least one gene is mutated). After mutation there are r different indicator strings. This operation is also executed for the s parent strings.

Step 5: Check of the restriction $\delta_{jk}^{xx} \leq \delta_j^x \delta_k^x$ (respectively their equipollent for categorical variables) and deletion of illegal interactions.

Step 6: The non-uniform mutation operator is applied to copies of parameter strings λ which correspond to the indicator strings δ. Here only smoothing parameters are mutated for which the value is 1. How many genes of a string are randomly selected and mutated is controlled by a random number (at least one gene is mutated). After mutation, there are r different parameter strings. This operation will also be executed for the s parent strings.

Step 7: The ABC operator is applied to the r indicator strings δ and thus generate r indicator offsprings. Apply the modIAC operator to the r parameter strings λ simultaneously and thus generate r parameter offsprings. Both crossover operators run independently.

Step 8: Check of the restriction $\delta_{jk}^{xx} \leq \delta_j^x \delta_k^x$ (respectively their equipollent for categorical variables) and deletion of illegal interactions.

Step 9: *Let r offsprings and s parent strings form the new population $P(t+1)$. Hence their are again $r+s$ indicator strings and $r+s$ parameter strings in the new population $(\tilde{\delta}, \tilde{\lambda})$.*

The selection in step $2, 3, 4, 6$ and 7 is implemented with respect to a probability distribution based on the strings' fitness (compare section 5.3.4).

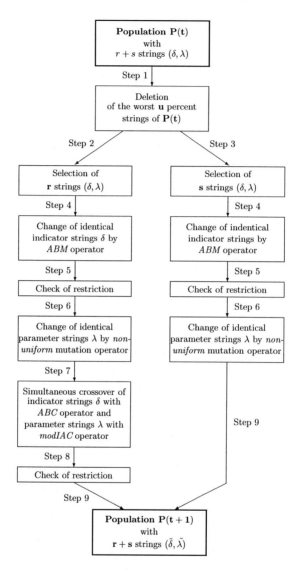

Figure 11.2. *The flowchart shows the structure of the modified selection procedure (modSP) which has been adapted to the problem of simultaneous selection of variables and smoothing parameters. Details in the text.*

Simultaneous Selection of Variables and Smoothing Parameters in Simulations

In the following sections we check the performance of variable selection and prediction accuracy of the genetic algorithm for simultaneous selection of variables and smoothing parameters and compare this approach with other methods in literature.

To evaluate the performance of the simultaneous genetic algorithm we have to check criteria $(1) - (4)$ as presented at the beginning of chapter 9.

This chapter is organized as follows: in the next section we analyse the effect of the two parameters crossover- and mutation probability of the genetic algorithm presented in chapter 11. Once more we check the effect to the results of the diverse information criteria described in chapter 4. This analysis of the problem of fine tuning is restricted to additive models, consisting of metrical covariates without interactions. More generally section 12.2 is extended to additive models which contain metrical and categorical covariates. Finally section 12.3 deals with the case of additive models consisting of metrical covariates as well as their interactions.

12.1 Effect of Fine Tuning in the Simultaneous Genetic Algorithm

For the genetic algorithm for simultaneous selection of variables and smoothing parameters the effect of parameters $p_c v$ (crossover probability for variable selection) and $p_m v$ (mutation probability for variable selection) is analysed.

Remark 24. In context with smoothing parameter determination in chapter 7 we have realized that the results have negligible differences for the diverse parameters of a genetic algorithm. This results have also been confirmed by simulations with the simultaneous genetic algorithm. Thus we do not dwell on this fact. We only deal closer with the parameters $p_c v$ and $p_m v$ which are introduced in sections 8.2.1 and 8.2.2.

To check the influence of $p_c v$ we simulate an additive model with 10 different functions $f_j(x_{ij}), j = 1 \ldots, 10$, where 5 functions are without effect, i.e. $f_j(x_{ij}) = 0$. The curves of the remaining functions are shown in Figure 6.5. We simulate 200 datasets, each one consisting of 200 independently and uniformly distributed observations with $\sigma = 0.2$. For estimation the single functions $f_j(x_{ij})$ are expanded in 20 cubic B-spline basis functions. As penalty we use the third difference of adjacent coefficients. The smoothing parameters are chosen from the interval $[10^{-4}, 10^4]$. The default parameters of the used genetic algorithm are: $popsize = 48$ strings, crossover probability (of the parameters) $p_c = 0.5$, mutation probability (of the variables) $p_m v = 0.5$, deletion of $u = 60$ percent of the worst strings, selection of $r = 28$ and $s = 20$ strings, $\nu = 0.5$, $T = 1000$ and $b = 1$. As information criteria we use the improved AIC.

Figure 12.1 shows the results of the genetic algorithm based on different choice of $p_c v \in \{0.1, 0.25, 0.5, 0.7, 1.0\}$. Here the top panel yields the number respectively portion of datasets with $0, 1, 2, \ldots$ misclassified variables. We see that the black coloured parts of the bars yield a comparable number of datasets with no incorrectly classified variable.

The left panel below yields the belonging boxplots which describe the prediction accuracy of the genetic algorithm with diverse crossover probabilities $p_c v$. Here we have no significant difference between the diverse crossover probabilities. A similar result can be recognized by the right panel below which shows the number of iterations up to the incidence of the results used in the evaluation. All boxplots need a comparable number of iterations up to convergence.

A similar simulation checks the effect of the mutation probability $p_m v$ on the performance of the genetic algorithm. Here we choose the probabilities from a set $p_m v \in \{0.1, 0.25, 0.5, 0.7, 1.0\}$. The crossover probability (of the variables) $p_c v$ is assumed as 0.5. The other default parameters are taken the same values as above. Concerning Figure 12.2 except for the case of $p_m v = 1.0$ all the other parameters show only very slight differences as to the number of datasets with misclassified variables. But in case of

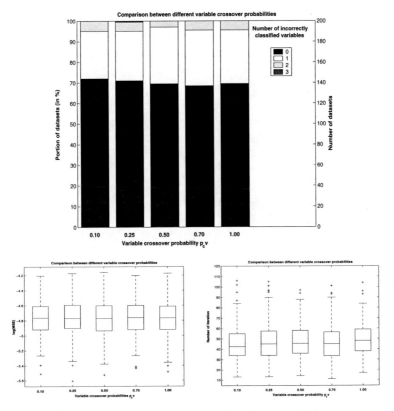

Figure 12.1. *Here the dependence of misclassification, prediction accuracy and simulation time on different crossover probability values $p_c v$ is shown.*

$p_m v = 1.0$ a significant lower number of datasets (approximately 10%) are completely correctly classified, i.e. each variable with effect is classified as variable with effect and vice versa.

The left panel below of Figure 12.2 describes the prediction accuracy for diverse mutation probabilities $p_m v$. Here we recognize a comparable performance of prediction accuracy for diverse probabilities. In regard to the convergence rate the right panel below presents the following tendency: in case of an increasing mutation probability $p_m v$ we have a decrease of the used iteration number in the datasets (also the variance of the used iteration numbers in the datasets decreases).

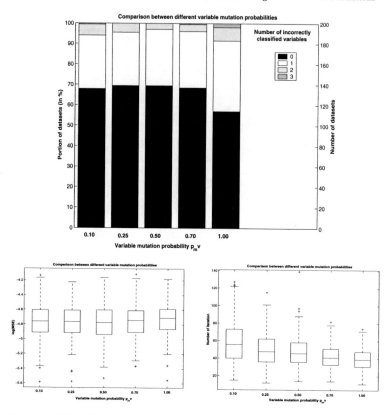

Figure 12.2. *Here the dependence of misclassification, prediction accuracy and simulation time on different mutation probability values $p_m v$ is shown.*

It should be noticed without any plots that in all datasets of the last two simulations the variables with effect are correctly classified. Thus all misclassifications are restricted to variables without any effect.

In chapter 9 diverse information criteria already produce significantly different results by application of the genetic algorithm for variable selection (compare Figure 9.1). In context with the simultaneous genetic algorithm we are also interested in analysing the performance of variable selection, prediction accuracy, and the rate of convergence for various information criteria. Again the top panel of Figure 12.3 shows the number respectively portion of datasets with $0, 1, 2, \ldots$ misclassified variables for AIC, improved AIC, BIC, GCV, and T criterion. The significantly best result

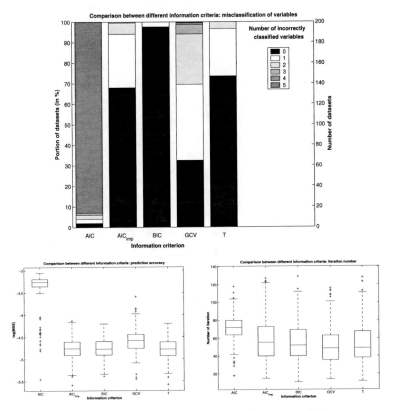

Figure 12.3. *Here the dependence of misclassification, prediction accuracy and simulation time on different information criteria is shown.*

yields the BIC: in 97% of the datasets all variables are correctly classified and the remaining datasets only contain one misclassified variable. Whereas improved AIC and the T criterion yield completely correctly classified variables in approximately 70% of the datasets, the GCV even in 32% of the datasets. Furthermore the genetic algorithm with one of the last three criteria contains datasets with more than one misclassified variable. AIC shows the worst result because almost each dataset contains 5 misclassified variables.

The left panel of Figure 12.3 below shows the prediction accuracy of the diverse information criteria. Here BIC, improved AIC and T criterion yield similar results. Compared with improved AIC and T criterion we realize that larger portion of completely correctly classified datasets of BIC do not lead to a significantly better prediction accuracy (at least in this simulation which contains relatively fewer variables). On the other hand compared with GCV we have a significant difference in prediction accuracy, because two-thirds of the datasets have at most one misclassification. The worst prediction accuracy is yielded by AIC. This fact is not surprising because of the large number of datasets ($> 90\%$), containing 5 misclassified variables.

Again the right panel of Figure 12.3 below presents the rate of convergence (i.e. the number of iterations up to the incidence of the result used in the evaluation) for the diverse information criteria. Except for AIC all other information criteria show comparable results for the number of iterations used.

Here we refer again to remark 22 where we mentioned that the error rate and the prediction accuracy strongly depend on the number of observations and the complexity of the model (i.e. the number of parameters to be estimated). Thus in the following we will frequently switch between BIC, AIC and improved AIC.

In the following simulations the additive model with 10 different functions $f_j(x_{ij}), j = 1, \ldots, 10$, where 5 functions have no effect (i.e. $f_j(x_{ij}) = 0$) is compared with other approaches in literature. As no common statistical software program contains a complete automatical procedure which simultaneously selects variables and smoothing parameter, we are restricted to software programs having separate tools which are applied successively.

Here we use software tools implemented in S-Plus and R:

- The software package S-Plus offers a restricted possibility of variable selection and simultaneous function estimation. First one calculates AIC for an initial model. Then one has to specify a list with other modelling alternatives. Each covariate can be dropped or integrated in the model as a linear term respectively as a cubic smoothing spline with a default smoothing parameter. Starting with the initial model the implemented function step successively calculates the AIC for all alternative models. If a current model yields a better AIC the previous model is replaced. Because of its implementation S-Plus can only run a relatively small number of different models. In the simulation each covariate is modelled linearly or as a cubic smoothing spline with degrees of freedom $df = 2, 6, 10, 14$. For further details see also sections 6.1.5 and 9.2.

- The statistic software package R offers the following approach to variable selection. The function stepAIC implemented in the package MASS chooses a model by AIC in a stepwise algorithm. This procedure is comparable with the step-function in S-Plus. But in R each covariate can be dropped or integrated in the current model as a linear term or as a polynomial up to degree 4. The user has the possibility to choose BIC as criterion. In the simulations below we have applied AIC and BIC. After variable selection the R-package mgcv (Wood (2000)) yields an automatic smoothing parameter selection based on a method first proposed by (Gu & Wahba (1991)). For further details also see section 6.1.6 and 9.2.

We simulate 200 datasets with $n_1 = 100$ respectively $n_2 = 200$ observations each one with noise $\sigma = 0.2$. The simulation with n_2 corresponds to that one used above in context with parameters' fine tuning. The default parameters of the used genetic algorithm are: $popsize = 38$ strings, crossover probability (of the variables) $p_c v = 0.25$, crossover probability (of the parameters) $p_c = 0.5$, mutation probability (of the variables) $p_m v = 0.1$, deletion of $u = 60$ percent of the worst strings, selection of $r = 28$ and $s = 10$ strings, $\nu = 0.5$, $T = 1000$ and $b = 1$.

As mentioned in Remark 22 BIC yields a penalization too rigid for datasets containing only a few observations as well as a large number of possible variables. In this cases application of AIC or improved AIC leads to more appropriate results. For the dataset with $n_1 = 100$ observations Figure 12.4 compares the number (respectively portion) of datasets with incorrectly classified variables in dependence of improved AIC and BIC. By use of the improved AIC in 86% of the datasets all variables are completely correctly classified, whereas, in no case, BIC shows correctly classified datasets. The improved AIC yields datasets with at most one incorrectly classified variable. But BIC also shows datasets with five misclassified variables. Hence Figure 12.4 advises to use the improved AIC for the dataset with $n_1 = 100$ observations as well as 10 possible variables (whereas each variable is expanded in 20 cubic B-splines).

For the dataset with $n_1 = 100$ observations Figure 12.5 yields the results of the comparison with the approaches in S-Plus and R (in these software programs we have used AIC). For genetic algorithm and stepwise procedures in S-Plus and R the top left panel shows the number (respectively portion) of datasets with $0, 1, 2, \ldots$ misclassified variables. The gentic al-

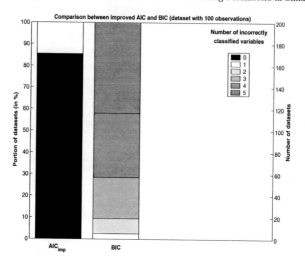

Figure 12.4. *Here the dependence of misclassification (dataset with $n_1 = 100$ observations) on improved AIC and BIC is shown.*

gorithm yields 86% of the datasets with no misclassified variable. S-Plus and R correctly determined 40% respectively 6% of the datasets, only. While the genetic algorithm shows datasets with at most one misclassified variable, S-Plus also has datasets with two incorrectly classified variables. The worst results are yielded by the stepwise procedure in R: here datasets containing four misclassified variables can occur. Furthermore most of the datasets (approximately 87%) have an error rate of three misclassified variables.

The top right panel of Figure 12.5 illustrates the number of datasets with errors in variables with effect (i.e. erroneously classified variables without effect). Here we realize that the genetic algorithm and the stepwise procedure in R correctly classify the variables with effect in all datasets. Also the stepwise procedure in S-Plus yields correct results in approximately 90% of the datasets. Thus the errors essentially occur in the variables without effect, i.e. more variables than necessary are included in the model.

The two panels below show the prediction accuracy of the diverse approaches. The left panel illustrates all three methods, i.e. the genetic algorithm and the stepwise procedures. The right panel is restricted to the genetic algorithm and the stepwise procedure in R. We realize that the

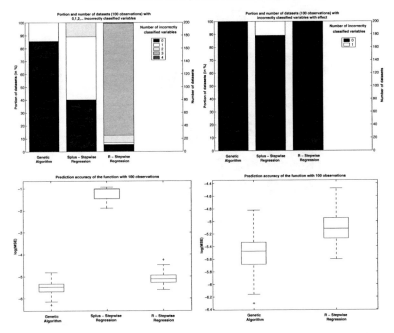

Figure 12.5. *The simulation results of the additive model with $n_1 = 100$ observations are shown here. The top panels yield the number (respectively portion) of datasets with incorrectly classified variables (the top left panel is restricted to the misclassified variables with effect). The panels below check the prediction accuracy by calculating $log(MSE)$.*

significantly worst estimation quality is given by the approach in S-Plus. This result depends on the limited choice of the models: as mentioned above each covariate is modelled linearly or as a cubic smoothing spline with degrees of freedom $df \in \{2, 6, 10, 14\}$.

The right panel suggests that there is also a significant difference in prediction accuracy between the genetic algorithm and the stepwise procedure in R. Summarizing we realize that the genetic algorithm outperforms the two stepwise procedures (in variable selection and prediction accuracy) in the dataset with $n_1 = 100$.

In the following dataset with $n_2 = 200$ observations we use BIC again. Figure 12.6 shows the respective results. In the top panel we significantly realize that S-Plus and the genetic algorithm approximately correctly clas-

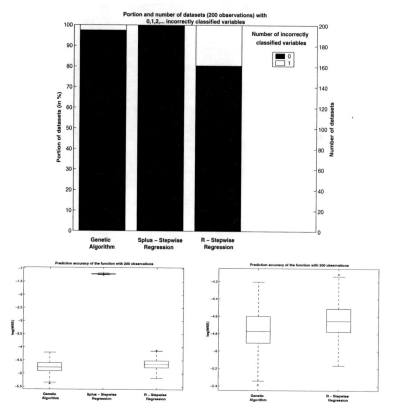

Figure 12.6. *The simulation results of the additive model (10 covariates, where 5 functions have no effect) with $n_2 = 200$ observations and noise $\sigma = 0.2$ are shown here. The top panel yields the number (respectively portion) of datasets with incorrectly specified variables. The two panels below show the prediction accuracy of the approaches. In the right panel the S-Plus approach is left out.*

sify all the datasets (only 2% of the datasets classified by the genetic algorithm contain one misclassified variable). The stepwise procedure in R yields 20% datasets with one misclassified variable.

If we analyse the prediction accuracy of the diverse approaches (compare the plots in Figure 12.6 below), the S-Plus approach shows the worst results as expected. The other two procedures yield significantly better results. But in this simulation the genetic algorithm also outperforms the stepwise procedure in R concerning prediction accuracy. It should be no-

ticed without any plot that in this simulation in each dataset all variables
with effect are correctly classified.

12.2 Additive Models with metrical and categorical Covariates

In general datasets contain metrical and categorical variables. For that
purpose we extend the additive model of section 12.1 to categorical variables. Here we simulate an additive model with 18 different functions: here
10 functions $f_j(x_{ij}), j = 1, \ldots, 10$, depend on metrical covariates (again 5
functions have no effect). Furthermore 8 functions $f_j(z_{ij}), j = 11 \ldots, 18$,
depend on binary covariates, where 5 functions have no effect, i.e. $f_j(z_{ij}) = 0$. We simulate 200 datasets each one consisting of 200 independently and
uniformly distributed obervations with $\sigma_1 = 0.2$ and $\sigma_2 = 0.4$.

For estimation the single functions $f_j(x_{ij})$ are expanded in 20 cubic B-
spline basis functions. As penalty we use the third difference of adjacent coefficients. The smoothing parameters are chosen from the interval $[10^{-4}, 10^4]$. The default parameters of the genetic algorithm used are:
$popsize = 38$ strings, crossover probability (of the variables) $p_c v = 0.25$,
crossover probability (of the parameters) $p_c = 0.5$, mutation probability
(of the variables) $p_m v = 0.1$, deletion of $u = 60$ percent of the worst
strings, selection of $r = 28$ and $s = 10$ strings, $\nu = 0.5$, $T = 1000$, and
$b = 1$. As information criterion we use BIC.

Figure 12.7 yields the results of the simulation with $n = 200$ observations
and $\sigma_1 = 0.2$. For comparison we use the software tools implemented in
S-Plus respectively R, as described in section 12.1. While the genetic algorithm and the stepwise procedure in S-Plus yield comparable results in
variable selection (top panel) the stepwise procedure in R leads to significantly worse error rates: 88% of the datasets have at most one misclassified variable. Hence there are fewer completely correctly classified
datasets. The stepwise procedure also contains datasets with up to five
misclassified variables. The other approaches yield at most one incorrectly
classified variable. It should be noticed without any plot that in this simulation in each dataset all variables with effect are correctly classified.
Thus the errors essentially occur in the variables without effect, i.e. more
variables than necessary are included in the model.

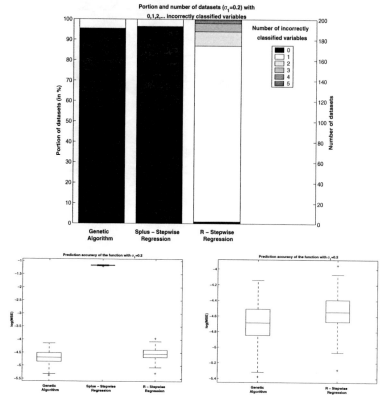

Figure 12.7. *The simulation results of the additive model with $n = 200$ observations and noise $\sigma_1 = 0.2$ are shown here. The top panel yields the number (respectively portion) of datasets with incorrectly specified variables. The two panels below show the prediction accuracy of the approaches. In the panel to the right the S-Plus approach is left out.*

The two plots of Figure 12.7 below show the prediction accuracy of the diverse approaches. Similar to section 12.1 S-Plus yields the worst results. Genetic algorithm and stepwise procedure in R lead to comparable estimations. But the right panel of Figure 12.7 below shows that the genetic algorithm significantly outperforms the procedure in R.

The Figure 12.8 shows the results for the similar simulation with $\sigma_2 = 0.4$. The panel to the right yields the number (respectively portion) of datsets with $0, 1, 2, \ldots$ incorrectly classified variables. While the genetic algorithm

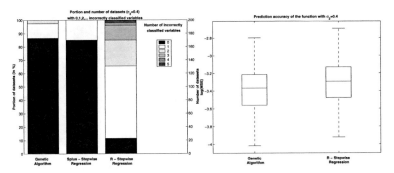

Figure 12.8. *The simulation results of the additive model with n = 200 observations and noise σ₂ = 0.4 are shown here. The left panel yields the number (respectively portion) of datasets with incorrectly specified variables. The right panel yields the prediction accuracy of the genetic algorithm and the stepwise procedure in R.*

and the stepwise procedure in S-Plus lead to comparable results (approximately 86% datasets have no misclassified variables) the approach in R yields significantly worse error rates: only 12% of the datasets have no misclassified variable. Furthermore the stepwise procedure in R contains datsets with up to five misclassified variables. The other two approaches contain at most one (S-Plus) respectively two (genetic algorithm) incorrectly classified variables. But it should be noticed again without any plot that in each dataset in this simulation all variables with effect are correctly classified.

The right panel of Figure 12.8 yields the prediction accuracy of the genetic algorithm and the stepwise procedure in R. Here we omit the illustration of the prediction accuracy of the S-Plus approach. Because of the restricted choice of models S-Plus again leads to similar worse results as received in the simulations above. The plot suggests that opposite to the stepwise procedure the genetic algorithm has lower $log(MSE)$-values and hence a more accurate estimation quality.

Remark 25. We have seen that the prediction accuracy of S-Plus leads to worse results. In the following simulation studies we will also omit the illustration of the prediction accuracy of S-Plus. Hence we restrict ourselves to a comparison between the genetic algorithm and the stepwise procedure in R. As the variable selection procedure in S-Plus does not depend on the choice of degrees of freedom S-Plus yields appropriate results in variable selection. Thus we will also consider this procedure of S-Plus in further simulation studies.

12.3 Additve Models with Interactions

In this section the additive model is extended by interactions between metrical respectively categorical variables.

First we analyse variable selection and prediction accuracy for an additive model consisting of metrical variables and their responding interactions. For that purpose we simulate an additive model with four metrical covariates $f_j(x_{ij}), j = 1, \ldots, 4$, where one function has no effect. Furthermore we have six interaction terms whereas four terms have no effect. We simulate 200 datasets each one consisting of 200 independently and uniformly distributed observations with $\sigma = 0.2$.

For estimation the single functions $f_j(x_{ij})$ are expanded in 15 cubic B-spline basis functions. For the interaction terms $f_{rs}(x_{ir}, x_{is})$ we choose two-dimensional cubic B-splines on a grid of 10 by 10 knots. In both cases the penalty is of third difference order. The smoothing parameters are chosen from the interval $[10^{-4}, 10^4]$. The default parameters of the genetic algorithm used are: *popsize* $= 32$ strings, crossover probability (of the variables) $p_c v = 0.25$, crossover probability (of the parameters) $p_c = 0.5$, mutation probability (of the variables) $p_m v = 0.5$, deletion of $u = 60$ percent of the worst strings, selection of $r = 22$ and $s = 10$ strings, $\nu = 0.5$, $T = 1000$, and $b = 1$. As information criterion we use the improved AIC.

Remark 26. In case of the interaction terms $f_{rs}(x_{ir}, x_{is})$ the R procedure uses thin plate regression splines with an user-defind number of knots (here we have chosen 15 knots). The S-Plus procedure bases on locally weighted regression smoothers. For further details about the choice of the default parameters see the manuals of R and S-Plus.

Figure 12.9 yields the results of the simulation. In comparison with variable selection we use the software tools implemented in S-Plus respectively R (with AIC), as described in section 12.1. From the top panel we realize that the genetic algorithm perfoms best: in approximately 80% of the

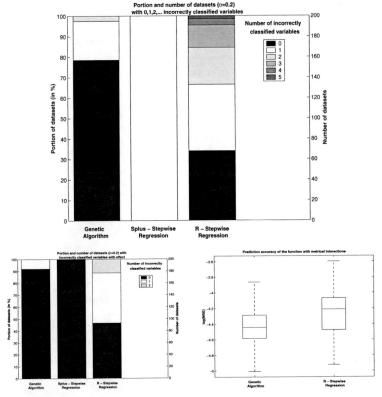

Figure 12.9. *The simulation results of the additive model with $n = 200$ observations and noise $\sigma = 0.2$ are shown here. The top panel yields the number (respectively portion) of datasets with incorrectly specified variables and the left panel below yields the respective results of incorrectly classified variables with effect . The right panel below shows the prediction accuracy of the genetic algorithm and the stepwise procedure in R.*

datasets we have no misclassification. Furthermore the genetic algorithm only yields a few datasets (3%) with 2 incorrectly classified variables. Otherwise the stepwise procedure in S-Plus shows one misclassified variable in each dataset. The worst results are generated by the stepwise procedure in R: here only 32% of the datasets are completely correctly classified and furthermore we find datasets, containing up to 5 incorrectly classified variables.

The left panel below of Figure 12.9 shows the results of misclassifications restricted to variables with effect. We realize that in all datasets the variables with effect were correctly classified by S-Plus. The genetic algorihm also yields similar results: in 8% of the datasets only one variable is misclassified. The approach in R shows larger problems in selection of the variables with effect, because in only 45% of the datasets all variables (with effect) are completely correctly classified.

The right panel below yields the prediction accuracy of the genetic algorithm and the stepwise procedure in R. It is obvious that the genetic algorithm has significantly better estimations compared with the approach in R.

In the last simulation study in this chapter we analyse an additive model consisting of 4 metrical and 4 categorical variables. Furthermore we have 6 interactions between metrical variables as well as 6 interactions between categorical variables. Altogether 8 variables respectively interactions have an effect (hence the other variables and interactions are without any effect). As default parameters we choose the same values as described in the last simulation.

Figure 12.10 yields the results for the additive model with interactions between metrical respectively categorical variables. In this simulation the S-Plus procedure also yields one misclassified variable in each dataset. In approximately 35% of the datasets the genetic algorithm shows no misclassified variable and in only 15% of the datasets we have more than 2 incorrectly classified variables. The procedure in R generates significantly worse results, because approximately 60% of the datasets have more than 2 misclassified variables. Furthermore in only 8% of the datasets the stepwise procedure in R leads to completely correctly classified datasets.

The left panel below shows comparable results for genetic algorithm and S-Plus procedure. Hence for these two approaches the errors essentially occur in the variables without effect, i.e. more variables than necessary are included in the model. In case of the stepwise procedure in R significantly more errors occur for variables with effect: only 40% datasets yield completely correctly classified variables with effect.

The right panel of Figure 12.10 below shows the prediction accuracy for the genetic algorithm and the procedure in R. Again we realize a significant difference between the two approaches. Furthermore the genetic algorithm

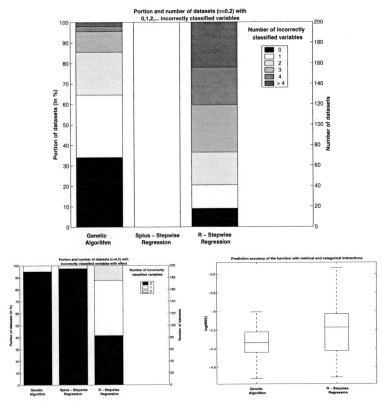

Figure 12.10.   *The simulation results of the additive model with $n = 200$ observations and noise $\sigma = 0.2$ are shown here. The top panel yields the number (respectively portion) of datasets with incorrectly specified variables and the left panel below yields the respective results of incorrectly classified variables with effect . The right panel below shows the prediction accuracy of the genetic algorithm and the stepwise procedure in R.*

shows a smaller variance in the estimations compared with the stepwise procedure in R.

We can summarize that in case of additive models with interactions the genetic algorithm also yields approriate results in variable selection and prediction accuracy and outperforms the two other approaches presented in most cases.

Application of the Simultaneous Genetic Algorithm to Rents for Flats

In the last years many large cities have published "rental guides" assisting renter respectively owner of flats to calculate their rents. Furthermore, according to the German rental law, owners are only allowed to increase the rents in dependence on "average rents" of comparable flats. To generally determine "average rents" several thousands of owners and renters are randomly chosen and interviewed in reference to the special equipment of the flat (e.g. bath equipment, kitchen, quality of heating or warm water system). Using further informations like e.g. rent, location of the flat or year of construction we have the possibility of determine the "average rent" (after specification of the respective criteria of the flat).

As basis of the statistical analysis in this chapter we have a random sample of 2055 flats from the census of the rental guide of the year 1998 in Munich. As response variable we choose

$y_i \equiv$ monthly net rent per square meters in Euro (this is calculated by the difference between the monthly rent and the estimated utility costs),

where $i = 1, \ldots, 2055$. Out of the approximately 200 variables of the original sample we use 3 metrical variables $x_{ij}, j = 1 \ldots, 3$ and 7 categorical variables $z_{ij}, j = 1, \ldots, 7$, as described in Table 13.1.

In context with this dataset we assume an additive model

$$
y_i = \beta_0 + \sum_{j=1}^{3} f_j(x_{ij}) + \mathbf{z}_i^T \boldsymbol{\alpha}_i + \sum_{r=1}^{2} \sum_{s=r+1}^{3} f_{rs}(x_{ir}, x_{is}) + \epsilon_i \quad ,
$$

where $\epsilon_i, i = 1, \ldots, 2055$, is independently and normally distributed.

The left panel of Figure 13.1 shows the normal probability plot for the

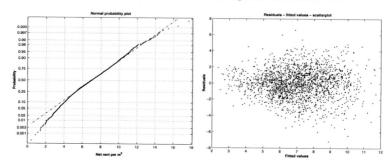

Figure 13.1. *Normal probability plot (left panel) and residual plot (right panel) for the sample of 2055 flats from the Munich rental guide of the year 1998.*

sample of the Munich rental guide. Here the data were plotted against a theoretical normal distribution. Departures from straight line indicate departures from normality. From the plot we realize that the center of the data shows approximately the same linear run as the theoretical curve. The tails (in particular the lower tail) show departures from the hypothetical straight line. As both ends of the normal probability plot bend below the hypothetical straight line the distribution from which the data were sampled is skewed to the left.

The right panel of Figure 13.1 shows a residual plot. Here the estimated residuals are plotted in dependence on the fitted values. We realize that the assumption of variance homogeneity is not valid.

Variable	Brief description	Scale
x_{i1}	floor space (in square meters)	metrical
x_{i2}	year of construction	metrical
x_{i3}	term of tenancy (in months)	metrical
z_{i1}	good location	binary
z_{i2}	best location	binary
z_{i3}	simple warm water supply	binary
z_{i4}	no warm water supply	binary
z_{i5}	no central heating	binary
z_{i6}	special auxiliary equipment in the bath	binary
z_{i7}	bath not tiled	binary

Table 13.1. *Used variables in the real dataset which basis on the rental guide of Munich (1998).*

To receive a selection of necessary variables respectively simultaneous estimation of the dataset we use the genetic algorithm for simultaneous selection of variables and parameters. Here the default parameters of the genetic algorithm are chosen as: $popsize = 32$ strings, crossover probability (of the variables) $p_c v = 0.25$, crossover probability (of the parameters) $p_c = 0.5$, mutation probability (of the variables) $p_m v = 0.5$, deletion of $u = 60$ percent of the worst strings, selection of $r = 22$ and $s = 10$ strings, $\nu = 0.5$, $T = 1000$, and $b = 1$.

The main effects of the 3 metrical variables are modelled by cubic B-splines with 20 knots; for the respective 3 interactions between metrical variables we choose two-dimensional cubic B-splines on a grid of 10 by 10 knots. In both cases the penalty is of third difference order. As model selection criterion we use improved AIC and BIC.

Remark 27. All interactions between categorical variables respectively categorical and metrical variables have been thrown off the model by the genetic algorithm and therefore shall not be mentioned anymore.

Concerning variable selection Table 13.2 shows the variables removed from the model by the genetic algorithm with improved AIC (respectively BIC). We realize that variables $x_{ij}, j = 1, \ldots, 3$ and $z_{ij}, j = 1, \ldots, 4, 6, 7$ are contained in the model for both cases. For the interaction term between x_{i1} and x_{i2} the two criteria lead to different results. Thus with the exception of the interaction between x_{i1} and x_{i2} the genetic algorithms with BIC and improved AIC yield the same model.

Figure 13.2 shows the effects of "floor space" and "year of construction" for the genetic algorithm with BIC respectively improved AIC. We realize that the main effect "floor space" shows comparable curves for both criteria:

Variable	BIC	improved AIC
z_{i5}	0	0
$x_{i1} \cdot x_{i2}$	0	1
$x_{i1} \cdot x_{i3}$	0	0
$x_{i2} \cdot x_{i3}$	0	0

Table 13.2. *Variables and interactions which are removed from the model by the genetic algorithm with improved AIC respectively BIC. Here value 0 means "the variable is not included in the model" and otherwise value 1 denotes the "inclusion of a variable in the model".*

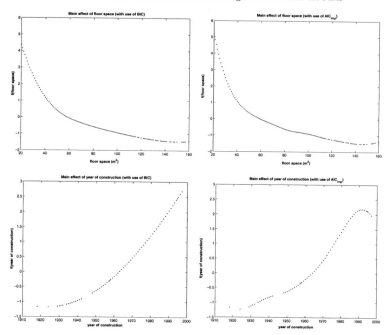

Figure 13.2. *Here the estimations of the main effects "floor space" and "year of construction" are shown. The plots on the left side yield the results for the genetic algorithm with BIC; the plots on the right side show the respective results for the genetic algorithm including the improved AIC.*

small flats are more expensive than larger ones but this effect becomes smaller with increasing floor space.

In case of the main effect "year of construction" the genetic algorithm also yields similar estimations for the two information criteria. The effect on the rents increases with more modern flats. Compared with flats before 1960 the effect on the rent is significantly larger for flats which have been built after 1960. With a year of construction later than 1990 there is a difference for flats between the respective curves of the genetic algorithm with BIC and improved AIC: while BIC yields further increase of the rent the improved AIC stabilizes the effect on the high level (respectively even decreases). This little difference in the run of the curves is given by the stronger penalization of the BIC. Comparable results can also be found in Lang & Brezger (2004).

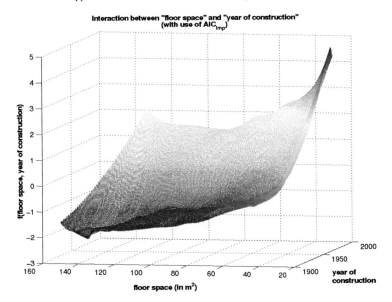

Figure 13.3. *Here is shown the effect of interaction between "floor space" and "year of construction" which is included in the model by application of the genetic algorithm with improved AIC.*

In Figure 13.3 we illustrate the effect of interaction between "floor space" and "year of construction" which is included in the model by application of the genetic algorithm with improved AIC. The plot shows that the monthly net rent per square meters has a significant dependence on floor space. However the monthly net rent depends slightly on the year of construction, only. Because the effect "year of construction" is relatively small we can understand that the BIC with its stronger penalization has not included this interaction in the model.

From the plot we realize that old flats built before 1940 with a floor space below $50m^2$ are cheaper than the average. Otherwise modern flats built in the year 1970 and later are more expansive than the average. The maximal rent have to be paid for small $(50m^2)$ respectively modern (year of construction: 1992) flat; otherwise large flats $(160m^2)$ built before World War II are the cheapest ones.

14

Conclusions

In this thesis various essential questions of statistics have been analysed by genetic algorithms, an approach which is uncommon and rarely used in statistics. Compared with many other approaches genetic algorithms have the advantage that they do not depend on knowledge or gradient information about the response surface; furthermore they offer the opportunity to independently jump out of a given local optimum. This property is interesting especially for users working with complex high-dimensional questions which have several local optima apart from the global optimum.

Understanding the exploitation-exploration-dilemma (section 5.3.1) is of fundamental interest to design an efficient genetic algorithm: the relative importance of exploration (i.e. exploring the search space and aquiring information about the nature of the space) and exploitation (i.e. local search near an optimum) varies with the run of the genetic algorithms' iteration process. Thus all genetic algorithms developed and presented in previous chapters aim to obtain an appropriate balance between exploitation and exploration during the whole iteration process. Our new genetic algorithms (e.g. GENcon, SMAD) include several adaptive respectively non-uniform operators specifically designed on basis of the dilemma. Compared with many other genetic algorithms including "classical" elements (see section 5.2.2) our algorithms lead to more accurate results in a shorter time.

The present thesis has used genetic algorithms to address specific statistical questions. These genetic algorithms however can be applied for diverse other (including non-statistical) fields. For example GENcon can be applied to those optimization problems which aim to minimize or maximize continous parameters (an example of mathematical optimization is given in appendix A).

The literature suggests various approaches to choose smoothing parameters (connected with penalized regression splines) appropriately, e.g. Hastie & Tibshirani (1990), Hastie, Tibshirani & Friedman (2001) and Lang & Brezger (2004). To our knowledge this is the first time genetic algorithms are used to address this problem. We performed diverse simulation studies to compare the quality of the genetic algorithm with other approaches in literature. The results of the gentic algorithm are of similar quality as most of the other methods described. In some cases the genetic algorithm offers advantages over other approaches.

We do not always have optimization problems of continous parameters but also of binary parameters. Especially the search for an appropriate model in context with subset selection is an important and common problem. In chapter 8 we introduced the genetic algorithm GENbin which is developed for optimization problems with binary parameters. We also compared GENbin with other approaches by using simulations with different complex additive models. The results yielded a distinctly better performance than the alternative software programs.

An interesting application is model selection in the analysis of gene expression data. In general the user has the problem of an extremely large number of variables (i.e. genes) and a small number of samples (e.g. the patient has a tumor respectively no tumor). The objective is to select only the relevant variables because in many datasets selected from the field of gene expression data most of the variables are irrelevant. In chapter 10 the genetic algorithm GENbin is adapted to this problem and compared with other common approaches. Here we have chosen a logistic model as an underlying model of the genetic algorithm.

In many statistical applications the user needs a simultaneous selection of variables and parameters. Examples are: variable selection with smoothing parameter choice or variable selection connected with simultaneous choice of knot locations. Unfortunately these complex problems of simultaneous selection of binary and continous parameters are not implemented in common software packages. In general software packages include separate tools which are applied successively.

For the first time this thesis presents an approach which enables a simultaneous selection of variables and parameters. The approach bases on a combination of genetic algorithms for continous and binary parmeters. In chapter 12 we have applied the genetic algorithm to additive models for which a simultaneous selection of variables and smoothing parameters is executed. Compared with software packages including separate tools in almost all simulations the genetic algorithm shows distinctly better results with respect to the error rate of the selected variables as well as prediction accuracy. Furthermore the application in a real dataset (Munich rental guide) confirmed the expected results.

Throughout the whole thesis we have always referred to the question which information criterion yields accurate results in a chosen dataset. This search for an appropriate information criterion is neglected in many publications and software tools; thus the user often receives suboptimal results. By diverse simulation studies we get the following result: various datasets are extremely dependent on the number of observations respectively the number of parameters included in the model. As an example we refer to variable selection: while, at the beginning of a selection procedure the model contains almost all variables this number decreases to a few at the end. Hence two extreme cases exist and information criteria with different penalization of model complexity can lead to different results: a strong penalization seems to be appropriate in case of many variables but otherwise can lead to insufficient results. Thus the user has to choose an appropriate information criterion which balances between the two extreme cases.

The approaches presented in this thesis can be extended and modified in different ways. All used algorithms base on genetic algorithms. An interesting alternative could be a mix between a genetic algorithm and another "fine tuning program", e.g. Newton method, local search, hill climbing or a simulated jumping method. Such a "fine tuning program" can be applied after a complete run of the genetic algorithm. Another possibility is a subsequent application of the two software tools in each iteration step of an algorithm. Diverse approaches in the field of low-dimensional mathematical optimization problems are already published by e.g. Areibi (2000), Fernandez & Amin (1997), Sekhon & Mebane (1998).

For function estimation we follow the concept of penalization of regression splines. Here each function is expanded in a generally large number of basis functions (in our case we use B-splines as basis functions). A possible overfit is prevented by a penalization term. Another possibility is the estimation of functions by adaptive selection of knots and hence the use of respective genetic algorithms for knot selection. First approaches in this direction are published by Pittman (2002). Furthermore a genetic algorithm for simultaneous selection of variables and knots could improve the results of chapter 12.

This thesis is restricted to additive models with normally distributed observations. Future research may apply genetic algorithms to other model classes respectively distribution families.

This thesis shows that genetic algorithms should be considered especially for complex and high-dimensional questions. A larger integration of genetic algorithms is suggested in particular for complex and very large datasets used in the statistical field.

A

The GENcon Package

Author. Rüdiger Krause <krause@stat.uni-muenchen.de>

Title. Genetic Algorithm for continuous values (GENcon)

The Software GENcon runs on MATLAB 6.

Description. The software tool GENcon was developed for solving optimization problems characterized by real values. Examples are the search of the minimum or maximum of a real-valued function or the calculation of smoothing parameters in context with penalized regression splines. GENcon is based on genetic algorithms and the basic theory of the algorithm used can be read in Krause & Tutz (2003).

Details. For application of GENcon the user has to specify the population matrix pop_init which has dimension $pop\_size \times M$. Here pop_size is the number of strings in the population and M yields the number of elements to be optimized. FIT_init is a vector of dimension $pop\_size \times 1$ and contains fitness values belonging to the respective strings of the population matrix. Fitness values are real numbers resulting from the chosen optimization function.

Example 8. Suppose we have a function $f(x)$ and search its minimum x_{min} in an interval $[a, b]$. Then $f(x_1)$ yields the fitness value of the variable x at location x_1.

The search space is characterized by two $(1 \times M)$-dimensional vectors ssl and ssr, specified by the user. Here ssl yields the minimum and ssr the maximum of the search space.

GENcon applies several built-in functions:

- *GENcon_sel* contains the selection procedure of the genetic algorithm;

- *GENcon_iac* contains the improved arithmetical crossover operator of the genetic algorithm;

- *GENcon_optifunction* contains the fitness function which has to be optimized.

For given arguments *GENcon_sel* and *GENcon_iac* run automatically. But *GENcon_optifunction* has to be specified by the user. The structure of this function is explained in example 9.

Arguments. The following parameters have to be specified by the user (arguments are presented in order of appearance in the genetic algorithm):

n	Dataset with number n. In general we have only one dataset, i.e. $n = 1$.
M	Number of elements to be optimized.
pc	Crossover probability with values in the interval $[0, 1]$.
ac	Weight vector characterizing the improved arithmetical crossover: $c = ac * a + (1 - ac)b, ac \in [0, 1]$ (compare formula 5.4 in section 5.3.2).
pop_size	Number of strings in a population. $\frac{pop\_size}{2}$ has always to be an even number.
rs	$rs \in [0, pop\_size]$ is the number of strings in a population selected for succeeding operations like crossover and mutation. rs has always to be an even number.
ds	$ds = round(a * \frac{pop\_size}{100})$ deletes the **a** percent worst strings of a population. **a** has to be chosen by the user.
t	Current number of iterations $t \in \{1, 2, \ldots, iter\}$.
iter	Maximal allowed number of iterations.
iter_num	Maximal number T of iterations (iter_num$\geq$iter).
bm	Corresponding system parameter b of the non-uniform mutation operator (compare section 5.3.3) with $bm \in \{1, 2, \ldots\}$.
value	System parameter for fitness calculation. The fitness values are always > 0, but the optimization function can also take values < 0. The transformation $fitness = (-1) * (optimization\ function - value)$ guarantees positive fitness. If GENcon terminates with the output change_value=[], the user has to choose a larger value > 0.
num	Number of strings to be saved.

termination	1: GENcon terminates after iter iterations;
	2: If there is no change of the best fitness value during equal iterations, GENcon is terminated.
equal	$equal \in \{1, 2, \ldots\}$.
result	1: Output of fitness values to be maximized;
	2: Output of true values of the optimization function.
ssl	$(1 \times M)$-vector with values in $\mathbb{R}$, where M is the number of variables which have to be optimized. ssl yields the minimum of the search space.
ssr	$(1 \times M)$-vector with values in $\mathbb{R}$. ssr yields the maximum of the search space.
pop_init	Matrix with values in $\mathbb{R}$ and dimension $pop\_size \times M$. Hence this population matrix has pop_size strings and each string has M elements which have to be optimized.
FIT_init	Vector with values in $\mathbb{R}$ and dimension $pop\_size \times 1$. FIT_init contains the fitness values belonging to the strings of the population matrix.

Values. The program GENcon returns the following arguments:

data_tFIT	Maximal fitness value of the last population before terminating GENcon.
data_tstr	String (with elements which have to be optimized) belonging to the fitness value data_tFIT.
data_btFIT	The largest fitness value taken during the whole run of GENcon.
data_btstr	String (with elements which have to be optimized) belonging to the fitness value data_btFIT.
data_tFIT_div	Fitness values of the num best strings of the last population before termination.
data_tstr_div	The num best strings corresponding to the fitness values of data_btFIT.
data_btFIT_div	The num best fitness values of the population containing data_btFIT.
data_btstr_div	The num best strings belonging to the fitness values data_btFIT_div.
itermem	Number of iterations before termination.

Depending on the argument **result** GENcon returns some elements (during running). If the user chooses $return = 1$, the output **Result** contains three columns: (i) iteration one to t, (ii) data_btFIT and (iii) data_tFIT. Otherwise $return = 2$ yields an output Result2 with three columns: (i) iteration one to t, (ii) $(-1) * (data\_btFIT - value)$ and (iii) $(-1) * (data\_tFIT - value)$.

Warnings.

- $\frac{pop\_size}{2}$ has always to be an even number.

- rs has always to be an even number.

- If GENcon terminates with output change_value=[ ], the user has to choose a bigger value > 0.

Example 9. As an example for practical use of the software package GENcon a mathematical optimization problem is considered now.

The *Rosenbrock's saddle* function (Jong (1975))

$$f(x_1, x_2) = \sum_{i=1}^{n-1} [100 \cdot (x_{i+1} - x_i^2)^2 + (x_i - 1)^2],$$

with $n = 2$, is a two-dimensional function optimized in the interval $[-2.048, 2.047]$. The global minimum of this function is at $f(1,1) = 0$ (compare Figure A.1). The objective of the optimization problem is the finding of the global optimum by GENcon. This apparently simple function implies a certain amount of risk to have a premature convergence in a local optimum.

The arguments are chosen as follows: n = 1, M = 2, pc = 0.5, pc = 0.5, ac = 0.5, pop_size = 48, rs = 28, ds = 29 (with a = 60), iter = 200, iter_num = 1000, bm = 1, value = 10000, num = 20, termination = 1, equal = 10, result = 2, ssl = $[-2.048 \; -2.048]$, ssr = $[2.047 \; 2.047]$. Furthermore the current iteration number goes from 1 to iter = 200. The population matrix pop_init has $pop\_size = 48$ rows and $M = 2$ columns and is generated by the MATLAB code

```
for i=1:M
    pophilf=ssl(1,i)+(ssr(1,i)-ssl(1,i))*rand(pop_size,1)
    if    i==1
        pop_init=pophilf
    else pop_init=[pop_init pophilf]
    end
end
```

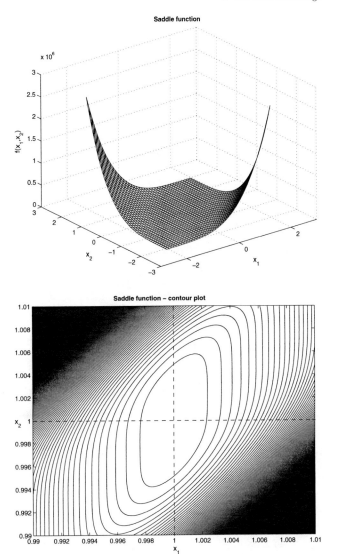

Figure A.1. *The plot at the top shows the Rosenbrock's saddle function and the plot below yields the contour plot of the function in the (x_1, x_2)-plain.*

To calculate the vector `FIT_init` consisting of fitness values we use the built-in function *GENcon_optifunction*. For the optimization problem described above, *GENcon_optifunction* consists of the Rosenbrock's saddel function as fitness function. Here the MATLAB code has the form

```
myFIT = (100*(wildcard(1,1)-(wildcard(1,2)).^2).^2
        + (wildcard(1,2)-1).^2),
```

where `wildcard(1,1)` and `wildcard(1,2)` correspond to the variables x_1 respectively x_2.

During running of GENcon, the output `Result2` yields three columns as described above. Hence for iterations $t = 1$ up to $t = 10$ we get e.g.

```
Result2 =
    1.0000   0.5806  0.5806
    2.0000   0.2740  0.2740
    3.0000   0.2740  0.6308
    4.0000   0.1835  0.1835
    5.0000   0.1719  0.1719
    6.0000   0.1390  0.1390
    7.0000   0.1390  0.1390
    8.0000   0.1390  0.1572
    9.0000   0.1390  0.1474
   10.0000   0.1331  0.1331
```

The first column yields the iterations up to t. The second column shows the transformed value of `data_btFIT`, i.e. $(-1) * (data\_btFIT - value)$, while the last column consists of the transformed values of `data_tFIT` calculated by $(-1) * (data\_tFIT - value)$.

After running GENcon returns the following arguments:

```
data_tFIT = 9.9999e+03
data_tstr = 1.6479 1.2768
data_btFIT = 1.0000e+04
data_btstr = 1.0889 1.0449
data_tFIT_div =
     1.0e+03 *
     Columns 1 through 6
     9.9999 9.9998 9.9998 9.9998 9.9998 9.9998 ...
```

```
data_tstr_div
    data_tstr_div(:,:,1) =
        1.6479 1.2768
    data_tstr_div(:,:,2) =
        2.0467 1.4345
    etc.
data_btFIT_div =
    1.0e+03 *
    Columns 1 through 6
    10.0000 9.9999 9.9998 9.9998 9.9998 9.9998 ...
data_btstr_div
    data_btstr_div(:,:,1) =
        1.0889 1.0449
    data_btstr_div(:,:,2) =
        1.7648 1.3235
    etc.
itermem = 191
```

For this simulated dataset GENcon yields values $x_1 = 1.0889$ and $x_2 = 1.0449$ for the position of the minimum. These results belong to the best fitness value data_btFIT = 1.0000^4. With transformation $(-1) *$ $(data\_btFIT - value)$ we get the true function value $f(x_1, x_2) = 0.0029$ which is close to the global minimum.

B

The SMAD Package

Author. Rüdiger Krause <krause@stat.uni-muenchen.de>

Title. SMoothing in Additive Models (SMAD)

The Software SMAD runs on MATLAB 6.

Description. With the software SMAD it is possible to calculate smoothing parameters of penalized basis functions (e.g. B-splines) in additive models with normally distributed response variable. The basic tool of SMAD is a genetic algorithm as described in Krause & Tutz (2003) (see also chapter 5) and implemented in the software package GENcon (appendix A).

Details. SMAD contains several tools, in fact

- *data tool* (specification of the used dataset),

- *basisfunction tool* (specification and calculation of basisfunctions),

- *initial population tool* (generation of the initial population for the succeeding genetic algorithm GENcon),

- *genetic algorithm tool* (contains the genetic algorithm basing on the software GENcon),

- *plot tool* (calculation and plot of main effects in additive models).

In each tool some parameters and variables have to be specified (for details see Arguments). It should be noticed, that the genetic algorithm tool contains built-in functions which are loaded into SMAD:

- *GENcon_sel* contains the selection procedure of the genetic algorithm;

- *SMADiac* contains the improved arithmetical crossover operator of the genetic algorithm;

- *optifunctool* contains the fitness function which has to be optimized.

Hence the user has to choose the path name where the subroutine is saved, appropriately.

In SMAD a `for`-loop for successive work of several different datasets is integrated. Thereby the parameter `dataset` yields the number of datasets. By different choices of the parameter `data` we can load datasets from file or simulate new datasets.

In case of simulated data we are often interested in the mean squared error (MSE). The MSE can be calculated by specification of the true function `f_i` in the data tool. If the user works with real datasets, the true function is generally unknown and the MSE can not be calculated. Thus in data tool a zero column ($f\_i = zeros(xdata, 1)$) have to be input to ignore the MSE-outputs.

Arguments. The following parameters have to be specified by the user (arguments are presented in order of appearance in the package SMAD):

xdata   Number of data for the independent variables.

M   Number of independent variables $x_1, \ldots, x_M$ in an additive model.

left   ($1 \times M$)-vector with values in $\mathbb{R}$. Each component i yields the minimum of the interval used for x_i.

factor   ($1 \times M$)-vector with values in $\mathbb{R}$. Each component i yields the width of the interval used for x_i.

> *Example 10.* For a vector $left = (-3, 0)$ and a vector $factor = (6, 1)$ the observations of variable x_1 are in the interval $[-3, 3]$. The observations of variable x_2 are in the interval $[0, 1]$.

sigma_hat Choice of standard deviation for noise $\epsilon \sim \mathcal{N}(0, sigma\_hat^2)$.

data   0: Dataset is loaded from file and the file name has to be specified by the user. In this case specifications of the parameter `sigma_hat` is not necessary;
           1: Simulation of new data.

innernodes Number of nodes in the interval $[left, (left + factor)]$.

bdeg   Degree of B-splines. In the package SMAD the default type of basis function are B-splines.

dord   Order of difference in the penalization matrix.

pop_size   Number of strings in the population. $\frac{pop\_size}{2}$ has always to be an even number.

search   1: Calculation of global smoothing parameters;
$ndx + bdeg - 1$: Calculation of local smoothing parameters, where $ndx = innernodes + 3$ is the total number of intervals between the nodes.

criterion 1: AIC information criterion;
2: Improved AIC information criterion;
3: BIC information criterion;
4: GCV information criterion;
5: T information criterion.

ssl $(1 \times search * M)$-vector with values in $\mathbb{R}$. ssl yields the minimum of the search space.

ssr $(1 \times search * M)$-vector with values in $\mathbb{R}$. ssr yields the maximum of the search space.

pop_init   Matrix with values in $\mathbb{R}$ and dimension $pop\_size \times search * M$. Hence this population matrix has pop_size strings and each string has $search * M$ variables which have to be optimized.

FIT_init   Vector with values in $\mathbb{R}$ and dimension $pop\_size \times 1$. FIT_init contains the fitness values belonging to the strings of the population matrix.

The genetic algorithm used is based on the package GENcon. The following parameters have to be specified: pc, ac, rs, ds, iter, iter_num, bm, value, num, termination, equal. Details of the parameters' choice and use can be found in the manual of GENcon (compare appendix A).

Values. The program GENcon returns the following arguments:

data_tFIT	Maximal fitness value of the last population before terminating GENcon.
data_tMSE	MSE which belongs to the fitness value data_tFIT.
data_tstr	String (with elements to be optimized) belonging to the fitness value data_tFIT.
data_tyhat	Estimator $\hat{y}$ belonging to the fitness value data_tFIT.
data_tstr	Weight vector $\hat{\beta}$ belonging to the fitness value data_tFIT.
data_btFIT	The biggest fitness value taken during the whole run of GENcon.
data_btMSE	MSE belonging to the fitness value data_btFIT.
data_btstr	String (with elements to be optimized) belonging to the fitness value data_btFIT.
data_btyhat	Estimator $\hat{y}$ belonging to the fitness value data_btFIT.
data_bthat	Weight vector $\hat{\beta}$ belonging to the fitness value data_btFIT.
data_tFIT_div	Fitness values of the num best strings of the last population before termination.
data_btFIT_div	The num best fitness values of the population containing data_btFIT.
itermem	Number of iterations before termination.
data_Bsch	Basis function matrix of the dataset.

Remark 28. Furthermore GENcon also yields the following arguments data_tMSE_div, data_tstr_div, data_tyhat_div, data_tbhat_div as well as data_btMSE_div, data_btstr_div, data_btyhat_ div, data_btbhat_div which have similar meaning as the arguments described above.

During running GENcon returns some elements: the output Result contains five columns: (i) iteration one to t, (ii) data_btFIT, (iii) data_btMSE, (iv) data_tFIT and (v) data_tMSE, each element up to iteration t.

Plots. The plot tool at the end of the software package SMAD can be activated by deletion of the symbol %. In case the user knows the true functions of the underlying main effects (in the additive model), this tool calculates and plots the estimators of the main effects. The following parameters have to be specified:

graphic 0: No plots;
    1: Plots.

start   Specification of the first dataset, for that the plot tool becomes applied.

dataset Specification of the last dataset, for that the plot tool becomes applied.

> *Example 11.* For *start* = 1 and *dataset* = 250 the plot tool is applied to dataset $n = 1$ up to 250. Choosing *start* = 10 and *dataset* = 10 the plot tool is only applied to dataset $n = 10$.

The plot tool yields the outputs:

y_main_plot For each dataset n the main effects' estimators are calculated. **y_main_plot** consists of n matrixes, each of dimension $M \times xdata$.

MSE_main MSE of the main effects for each dataset. **MSE_main** has dimension $n \times M$.

Remark 29. If *graphic* = 1 the user also gets boxplots of $log(MSE\_main)$ of the main effects and $log(data\_btMSE)$ (compare e.g. Figures 6.6 and 6.7 in section 6.3).

Warnings.

- If the number of datasets $n > 1$, the parameters M and xdata have to be equal for each dataset.

- $\frac{pop\_size}{2}$ has always to be an even number.

- rs has always be an even number.

- If GENcon terminates with output change_value=[ ], the user has to choose a bigger value > 0.

C

The GENbin Package

Author. Rüdiger Krause <krause@stat.uni-muenchen.de>

Title. Genetic Algorithm for binary values (GENbin)

The Software GENbin runs on MATLAB 6.

Description. The software tool GENbin was developed for solving optimization problems characterized by binary values. Examples are variable selection or knot selection at predefined knot locations. GENbin is based on genetic algorithms and the basic theory of the used algorithm can be read in chapter 8.

Details. For application of GENbin the user has to specify the population matrix pop which has dimension $pop\_size \times met * intm$. Here pop_size is the number of strings in the population and met yields the number of elements to be optimized. The parameter intm is calculated automatically and shows the number of interactions between two different variables (this procedure is useful e.g. for variable selection). In cases there are no interactions between two variables (e.g. for knot selection) the user must specify intm with 0 and thus the population matrix has dimension $pop\_size \times met$.

FIT is a vector of dimension $pop\_size \times 1$ and contains fitness values belonging to the respective strings of the population matrix. Fitness values are real numbers resulting from the chosen optimization function (e.g. AIC criterion).

Moreover specification of an iteration index t which yields the current value of the iteration necessary is also necessary. Generally the user programs a for-loop of the form "for t = 1 to iter_num", where iter_num is the maximal number of iterations.

GENbin applies several built-in functions:

- *GENbin_popinit* contains a program for calculating a random initial population;

- *GENbin_sel* contains the selection procedure of the genetic algorithm;

- *GENbin_abc* contains the adaptive binary crossover operator of the genetic algorithm.

For given arguments *GENbin_sel* and *GENbin_abc* run automatically. *GENbin_popinit* can optionally be used for designing an initial population (compare Remark 30).

Remark 30. If the user has to design an initial population, he can choose the built-in function *GENbin_popinit*. The population matrix pop has dimension $pop\_size \times met * intm$, where pop_size, met respectively intm have to be specified by the user (compare also Details). In cases there are interactions between variables (i.e. intm > 0) an integrated program checks the restriction (8.1) that an interaction can only be included if both variables $\mathbf{x}_j$ and $\mathbf{x}_k$ are included.

Arguments. The following parameters have to be specified by the user (arguments are presented in order of appearance in the genetic algorithm):

pop_size Number of strings in a population. $\frac{pop\_size}{2}$ has always to be an even number.

met   Number of variables or parameters which have to be optimized.

intm $(met-1)*met)/2$: Number of interactions between two variables; 0: No interactions between variables.

pm   Mutation probability of the adaptive binary mutation operator (formula (8.3)).

pc   Crossover probability of the adaptive binary crossover operator (formula (8.2)).

rs $rs \in [0, pop\_size]$ is the number of strings in a population selected for succeeding operations like crossover and mutation. rs has always to be an even number.

ds $ds = round(a * \frac{pop\_size}{100})$ deletes the a percent worst strings of a population. a has to be chosen by the user.

t   Current number of iterations $t \in \{1, 2, \ldots, iter\}$.

iter   Maximal allowed number of iterations.

iter_num Maximal number T of iterations (iter_num$\geq$iter).

bm   Corresponding system parameter b of the non-uniform crossover and mutation operators with $bm \in \{1, 2, \ldots\}$.

minmax 1: The users' fitness criterion (or optimization function) is minimized (e.g. AIC, GCV);
        2: The users' fitness criterion (or optimization function) is maximized.

value   System parameter for fitness calculation. The fitness values are always > 0, but the optimization function can also take values < 0. The transformation $fitness = (-1) * (optimization\ function - value)$ guarantees positive fitness. If GENbin terminates with the output change_value=[ ], the user has to choose a larger value > 0. t and FIT have to be specified by the user. pop can also be designed by the built-in function *GENbin_popinit*.

Values. The program GENbin returns the following argument:

pop   A new population matrix with dimension $pop\_size \times met * intm$.

Warnings.

- $\frac{pop\_size}{2}$ has always to be an even number.

- rs has always to be an even number.

- If GENbin terminates with output change_value=[ ], the user has to choose a bigger value > 0.

References

Akaike, H. (1973). A new look at statistical model identification. *IEEE Transactions on Automatic Control* **19**, 716–723.

Alon, U., Barkai, N., Notterman, D., Gish, K., Ybarra, S., Mack, D., and Levine, A. (1999). Broad patterns of gene expression revealed by clustering analysis of tumor and normal colon tissues probed by oligonucleotide arrays. *Proceedings of National Academy of Science, Cell Biology* **96**, 6745–6750.

Antonisse, J. (1989). A new interpretation of schema notation that overturns the binary coding constraint. In J. Schaffer (Ed.), *Proceedings of the Third International Conference on Genetic Algorithms*. San Mateo: Morgan Kaufman Publishers.

Areibi, S. (2000). An integrated genetic algorithm with dynamic hill climbing for vlsi circuit partitioning. In A. A. Freitas, W. Hart, N. Krasnogor, & J. Smith (Eds.), *Data Mining with Evolutionary Algorithms*, pp. 97–102. Las Vegas, Nevada, USA.

Baker, J. (1985). Adaptive selection methods for genetic algorithm. In J. Greffenstette (Ed.), *Proceedings of the First International Conference on Genetic Algorithms*, pp. 101–111. Hillsdale, NJ: Lawrence Erlbaum Associates.

Baker, J. (1987). Reducing bias and inefficiency in the selection algorithm. In J. Greffenstette (Ed.), *Proceedings of the Second International Conference on Genetic Algorithms*, pp. 14–21. Hillsdale, NJ: Lawrence Erlbaum Associates.

Bethke, A. (1980). *Genetic Algorithms as Function Optimizers*. Ann Arbor, MI: PhD tesis, University of Michigan.

Bhandari, D., Murthy, C., and Pal, S. (1996). Genetic algorithm with elitist model and its convergence. *International Journal of Pattern Recognition and Artificial Intelligence* **10**, 731–747.

Biller, C. and Fahrmeir, L. (2002). Bayesian varying-coefficient models using adaptive regression splines. *Statistical Modelling* **11**, 1–17.

Bishop, C. M. (1995). *Neural Networks for Pattern Recognition.* Oxford: Clarendon Press.

de Boor, C. (1978). *A Practical Guide to Splines.* New York, Heidelberg, Berlin: Springer.

de Boor, C. (1993). B(asic)-spline basics. In L. Piegel (Ed.), *Fundamental Developments of Computer-Aided Geometric Modeling*, pp. 27–49. London: Academic Press.

Breiman, L., Friedman, J. H., Olshen, R. A., and Stone, C. J. (1984). *Classi cation and Regression Trees.* Pacific Grove: Wadsworth and Brooks/Cole.

Craven, P. and Whaba, G. (1979). Smoothing noisy data with spline functions. *Numerische Mathematik* **31**, 377–403.

Darwin, C. (1859). *The origin of species by means of natural selection, or the preservation of favoured races in the struggle for life.* London: Penguin Books.

Davis, L. (1991). *Handbook of Genetic Algorithms.* New York: Van Nostrand Reinhold.

Deb, K. and Goldberg, D. E. (1991). A comparative analysis of selection schemes used in genetic algorithms. In G. J. Rawlins (Ed.), *Foundations of Genetic Algorithms*, pp. 69–93. San Mateo, CA: Morgan Kaufmann Publishers.

Dierckx, P. (1995). *Curve and Surface Fitting with Splines.* Oxford: Clarendon Press.

Dixmier, J. (1984). *General Topology.* New York, Berlin: Springer-Verlag.

Eiben, A., Arts, E., and Van Hee, K. (1991). Global convergence of genetic algorithms: On infinite markov chain analysis. In H.-P. Schwefel & R. Männer (Eds.), *Proceedings of the First International Conference on Parallel Problem Solving from Nature (PPSN), Lecture Notes in Computer Science, Vol. 496*, pp. 4–12. Heidelberg: Springer.

Eilers, P. H. C. and Marx, B. D. (1996). Flexible smoothing with b-splines and penalties. *Stat. Science 11*(2), 89–121.

Eshelman, L. and Schaffer, J. (1993). Real-coded genetic algorithms and interval-schemata. In L. D. Whitley (Ed.), *Foundations of Genetic Algorithms 2*. San Mateo: Morgan Kaufman Publishers.

Fahrmeir, L. and Tutz, G. (2001). *Multivariate statistical modelling based on generalized linear models, 2nd edition*. New York: Springer Verlag.

Fernandez, J. and Amin, S. (1997). Simulated jumping in genetic algorithms for a set of test functions. In *Proceedings of the IASTED International Conference on Intelligent Information Systems*. Grand Bahama Island.

Fogel, D. (1992). *Evolving Artificial Intelligence. Doctoral Dissertation.* San Diego: University of California.

Fogel, L. J., Owens, A. J., and Walsh, M. J. (1965). Artificial intelligence through a simulation of evolution. In F. Maxfield, Callahan (Ed.), *Biophysics and cybernetic systems*. Washington: Spartan.

Friedman, J. (1991). Multivariate adaptive regression splines (with discussion). *Annals of Statistics 19*(1), 1–141.

Friedman, J., Hastie, T., and Tibshirani, R. (2000). Additive logistic regression: A statistical view of boosting (with discussion). *The Annals of Statistics 28*(2), 337–407.

Goldberg, D. and Segrest, P. (1987). Finite markov chain analysis of genetic algorithms. In J. Grefenstette (Ed.), *Proceedings of the Second International Conference on Genetic Algorithms*. Hillsdale, NJ: Lawrence Erlbaum Associates.

Goldberg, D. E. (1989). *Genetic Algorithms in Search, Optimization and Machine Learning*. Reading, MA: Addison-Wesley.

Goldberg, D. E. (1991). Real-coded genetic algorithms, virtual alphabets, and blocking. *Complex Systems 5*, 139–167.

Goldberg, D. E., Deb, K., and Korb, B. (1991). Do not worry, be messy. In R. Belew & L. Booker (Eds.), *Proceedings of the Fourth International Conference on Genetic Algorithms*, pp. 24–30. San Mateo, CA: Morgan Kaufmann Publishers.

Gu, C. and Wahba, G. (1991). Minimizing gcv/gml scores with multiple smoothing parameters via the newton methods. *SIAM Journal of Scientific and Statistical Computing 12*, 383–398.

Hamadeh, H. and Afshari, C. A. (2000). Gene chips and functional genomics. *American Scientist 88*, 508–515.

Hart, W. and Belew, R. (1991). Optimizing an arbitrary function is hard for the genetic algorithm. In R. Belew & L. Booker (Eds.), *Proceedings of the Fourth International Conference on Genetic Algorithms*, pp. 190–195. San Mateo CA: Morgan Kaufmann Publishers.

Hartl, R. (1990). *A global Convergence Proof for a Class of Genetic Algorithms. Technical Report.* Wien: Technische Universität Wien.

Hastie, T. and Tibshirani, R. J. (1990). *Generalized Additive Models.* London: Chapman and Hall.

Hastie, T. and Tibshirani, R. J. (1993). Varying-coefficient models. *Journal of the Royal Statistical Society B* **55**, 757–796.

Hastie, T., Tibshirani, R. J., and Friedman, J. (2001). *The Elements of Statistical Learning.* New York: Springer.

Haupt, R. L. and Haupt, S. E. (1998). *Practical Genetic Algorithms.* New York: Wiley.

Herrera, F., Lozano, M., and Verdegay, J. L. (1998). Tackling real-coded genetic algorithms: Operators and tools for behavioural analysis. *Artificial Intelligence Review 12*(4), 265–319.

Holland, J. (1975). *Adaption in neural and artificial systems.* Ann Arbor: University of Michigan Press.

Hurvich, C. and Simonoff, J. (1998). Smoothing parameter selection in nonparametric regression using an improved akaike information criterion. *Journal of the Royal Statistical Society B 60*(2), 271–293.

Jong, K. (1975). *Analysis of the Behavior of a Class of Genetic Adaptive Systems. PhD. Thesis.* Ann Arbor: University of Michigan.

Kirkpatrik, S., Gelatt, Jr., C. D., and Vecchi, M. P. (1983). Optimization by simulated annealing. *Science* **220**, 671–680.

Krause, R. and Tutz, G. (2003). Additive modelling with penalized regression splines and genetic algorithms. *Discussion Paper 312, SFB 386, University of Munich.*

Krause, R. and Tutz, G. (2004). Simultaneous selection of variables and smoothing parameters in additive models. In D. Baier & K.-D. Wernecke (Eds.), *Proceedings of the 27th Annual GfKl Conference, University of Cottbus.* Heidelberg-Berlin: Springer-Verlag (In preparation).

Lang, S. and Brezger, A. (2004). Bayesian p-splines. *Journal of Computational and Graphical Statistics* **13**, 183–212.

Leitenstorfer, F. (2003). *Robustifizierte Schätzung im binären Regressionsmodell.* Diplomarbeit, University of Munich.

Louis, S. and Rawlins, G. (1992). Predicting convergence time for genetic algorithms. *Technical report, Department of Computer Science, Indiana University, Bloomington, IN, 47405*.

Michalewicz, Z. (1996). *Genetic Algorithms + Data Structures = Evolution Programs*. Berlin, Heidelberg: Springer.

Miller, A. (2002). *Subset Selection in Regression*. Boca Raton, London, New York: Chapman & Hall/CRC.

Mitchell, M. (1996). *An Introduction to Genetic Algorithms*. Cambridge, Massachusetts: MIT Press.

Mühlenbein, H. and Schlierkamp-Voosen, D. (1993). Predictive models for the breeder genetic algorithm 1: Continous parameter optimization. *Evolutionary Computation 1*(1), 25–49.

Oliveira, L. S., Benahmed, N., Sabourin, R., Bortolozzi, F., and Suen, C. Y. (2001). Feature subset selection using genetic algorithms for handwritten digit recognition. In *Proceedings of the 14th Brazilian Symposium on Computer Graphics and Image Processing*, pp. 362–369. Floriano'polis-Brazil: IEEE Computer Society.

Parise, H., Wand, M. P., Ruppert, D., and Ryan, L. (2001). Incorporation of historical controls using semiparametric mixed models. *Applied Statistics 50*(1), 31–42.

Pittman, J. (2002). Adaptive splines and genetic algorithms. *Journal of Computational and Graphical Statistics 11*(3), 1–24.

Radcliffe, N. (1991a). Equivalence class analysis of genetic algorithms. *Complex Systems 5*(2), 183–205.

Radcliffe, N. (1991b). Forma analysis and random respectful recombination. In R. Belew & L. Booker (Eds.), *Proceedings of the Fourth International Conference on Genetic Algorithms*, pp. 222–229. San Mateo, CA: Morgan Kaufman.

Rechenberg, I. (1973). *Evolutionsstrategie - Optimierung technischer Systeme nach den Prinzipen der biologischen Evolution*. Stuttgart: Frommann - Holzboog.

Rice, J. (1984). Bandwidth selection: plug in methods versus classical methods. *Annals of Statistics 12*, 1215–1230.

Rudolph, G. (1994). Convergence analysis of canonical genetic algorithms. *IEEE Transactions on Neural Networks, special issue on evolutionary computation 5*(1), 96–101.

Ruppert, D. and Carroll, R. (2000). Spatially-adaptive penalties for spline fitting. *Australian and New Zealand Journal of Statistics 42*(2), 205–223.

Ruppert, D., Wand, M., and Carroll, R. (2003). *Semiparametric Regression.* Cambridge: Cambridge University Press.

Schwarz, G. (1978). Estimating the dimensions of a model. *Annals of Statistics* **6**, 461–464.

Schwefel, H.-P. (1975). *Evolutionsstrategie und numerische Optimierung.* Berlin: Dissertation an der Technischen Universität Berlin, Abteilung für Prozessautomatisierung.

Schwefel, H.-P. (1995). *Evolution and Optimum Seeking.* New York: John Wiley & Sons.

Sekhon, J. S. and Mebane, W. R. (1998). Genetic optimization using derivatives: Theory and application to nonlinear models. *Political Analysis* **7**, 187–210.

Tipping, M. E. (2000). The relevance vector machine. In T. L. S.A. Solla & K.-R. Müller (Eds.), *In Advances in Neural Information Processing Systems.* Cambridge, MA: MIT Press.

Tipping, M. E. (2001). Sparse bayesian learning and the relevance vector machine. *Journal of Machine Learning Research* **1**, 211–244.

Tutz, G. (2003). Response smoothing estimators in binary regression. *Discussion Paper 312, SFB 386, University of Munich.*

Voigt, H. and Anheyer, T. (1994). Modal mutations in evolutionary algorithms. *Proc. of the First IEEE International Conference on Evolutionary Computation* **1**, 88–92.

Vose, M. (1991). Generalizing the notation of schema in genetic algorithms. *Artificial Intelligence* **50**, 385–396.

Vose, M. and Liepins, G. (1991). Schema disruption. In R. Belew & L. Booker (Eds.), *Proceedings of the Fourth International Conference on Genetic Algorithms*, pp. 237–242. San Mateo, CA: Morgan Kaufman.

Wallet, B. C., Marchette, D. J., Solka, J. L., and Wegman, E. J. (1996). A genetic algorithm for best subset selection in linear regression. In *Proceedings of the 28th Symposium on the Interface.*

Whitley, D. (1989). The genitor algorithm and selection pressure: Why rank-based allocation of reproductive trials is best. In J. Schaffer (Ed.), *Proceedings of the Third International Conference on Genetic*

Algorithms, pp. 116–121. San Mateo, CA: Morgan Kaufmann Publishers.

Wood, S. (2000). Modelling and smoothing parameter estimation with multiple quadratic penalties. *Journal of the Royal Statistical Society B 62*(2), 413–428.

Wood, S. (2001). mgcv: Gams and generalized ridge regression for R. *R News 1*(2), 20–25.

Wright, A. (1991). Genetic algorithms for real parameter optimization. In G. Rawlin (Ed.), *Foundations of Genetic Algorithms 1*. San Mateo: Morgan Kaufman Publishers.

Yang, J. and Honavar, V. (1997). Feature subset selection using a genetic algorithm. *IEEE Intelligent Systems 13*, 44–49.

Index

CURRICULUM VITA

Personal Data

Name	Krause, *Rüdiger* Ernst
Place of Birth	Kaiserslautern
Nationality	German

Education

10/1993 - 02/1999	Studium Lehramt für Gymnasien in den Fächern Mathematik, Physik und Erziehungswissenschaften, Technische Universität Kaiserslautern
02/1999 - 10/1999	Studium Diplom-Mathematik mit Nebenfach Physik, Technische Universität Kaiserslautern
02/1999 - 10/1999	Diploma Thesis „Die Bedeutung der Fehlspezifikation innerhalb neuronaler Netze" Supervisor: Prof. Dr. J. Franke Technische Universität Kaiserslautern
seit 10/1999	Wissenschaftlicher Mitarbeiter am Institut für Statistik und im Sonderforschungsbereich 386 „Statistical Analysis of Discrete Structures" (Project „Locally adaptive Approaches") der Ludwig-Maximilians-Universität München
05/2004	Doctoral Degree: Doktor der Naturwissenschaften (Dr. rer. nat.) Ludwig-Maximilians-Universität München